Regional Perspectives in Archaeology

From Strategy to Narrative

Richard Tabor

BAR International Series 1203
2004

Published in 2019 by
BAR Publishing, Oxford

BAR International Series 1203

Regional Perspectives in Archaeology

© Richard Tabor and the Publisher 2004

The author's moral rights under the 1988 UK Copyright,
Designs and Patents Act are hereby expressly asserted.

All rights reserved. No part of this work may be copied, reproduced, stored,
sold, distributed, scanned, saved in any form of digital format or transmitted
in any form digitally, without the written permission of the Publisher.

ISBN 9781841713502 paperback
ISBN 9781407326139 e-book

DOI https://doi.org/10.30861/9781841713502

A catalogue record for this book is available from the British Library

This book is available at www.barpublishing.com

BAR Publishing is the trading name of British Archaeological Reports (Oxford) Ltd.
British Archaeological Reports was first incorporated in 1974 to publish the BAR
Series, International and British. In 1992 Hadrian Books Ltd became part of the BAR
group. This volume was originally published by John and Erica Hedges in conjunction
with British Archaeological Reports (Oxford) Ltd / Hadrian Books Ltd, the Series
principal publisher, in 2004. This present volume is published by BAR Publishing,
2019.

BAR titles are available from:

        BAR Publishing
        122 Banbury Rd, Oxford, OX2 7BP, UK
EMAIL   info@barpublishing.com
PHONE +44 (0)1865 310431
FAX   +44 (0)1865 316916
        www.barpublishing.com

Foreword

This book is an updated version of my Ph D thesis, presented in November 1998. Its exploration of the relationship between regional survey sampling and narrative arose from my need to formulate a coherent strategy for the South Cadbury Environs Project, of which I continue to be the fieldwork director.

The thesis considered the extent to which regional narratives derived from archaeological survey are conditioned by techniques and sampling strategies. Its principal sources are reports of fieldwork in Africa, north and south America and Europe, as well as recent technical and theoretical publications. Recognising the hermeneutical process which informs every stage of the archaeologist's work, I have attempted to trace through the transformation of data towards higher social and ideological accounts.

A means for conceptualising archaeological space on a regional scale has been constructed to optimise the efficacy of the choice and distribution of techniques and the subsequent analysis of data. To this end aspects of several surveys are scrutinised. This culminates in a method for classifying surveys according to (1) their distribution of resources; (2) their spatial resolution; and (3) their chronological resolution.

Although recently completed and ongoing surveys are included it is important to bear in mind the often prolonged delay between the completion of fieldwork and publication. New techniques are evolving, old ones are being refined and improved, particularly in the area of remote sensing, which has been accorded a larger role here than in other books on survey.

Acknowledgements

The inception of my thesis, and hence this book, was entirely due to the creation of the South Cadbury Environs Project, prompted in particular by two of the University of Glasgow team which, in the early 1990s, was working on a report of Leslie Alcock's excavations of Cadbury Castle, Somerset. It was Phil Freeman who, in a state of mutual mild inebriation, suggested the need for fieldwork in a landscape in which there had been no systematic research, other than on the hilltop. And his colleague, Paul Johnson, then gave weeks of his free time over several years doing geophysical survey and teaching it to volunteers, and becoming a founding co-director of the project. Paul's work remains the biggest single influence over my project design. Subsequently Peter Leach (then of Birmingham University field Archaeology Unit) initiated what is an ongoing programme of excavations for the project, lending the weight of his department at a crucial time, and drawing me to Birmingham to do my Ph D.

My thesis benefited from discussions with Susan Limbrey, Ann Woodward and particularly Vince Gaffney, whose own work I regard as inspirational. Inspiration of a quite different kind came from my former Philosophy teacher, Professor Frank Cioffi of the University of Essex, who made academic enquiry seem worthwhile. Since 2001 funding from the Leverhulme Trust has allowed me to be a full time researcher for the South Cadbury Environs Project as a Fellow of the University of Bristol. Whilst there I have received great support from my colleagues Mark Corney and particularly Michael Costen. The fruits are in some of what has been added to the original thesis.

None of this would have been possible without the love, encouragement and support of my partner, Mandy Tabor. She turned her life upside down, working full time, caring for Martha, Barnaby and Oscar (born early in my third year of study) to accommodate my studies. And the children showed great forbearance in allowing Dad to spend so much time in seclusion so that he might one day do the job he really wanted to do!

Lastly, I owe more than I can say to Professor Leslie Alcock, who allowed a ten year boy to begin living a dream - and to Cadbury Castle which is still part of that dreamlife.

Richard Tabor, December 2003

Contents

1: Regional Philosophies — 1
- 1.1 Introduction — 1
- 1.2 Terms of reference — 1

2: Landscape Narratives — 4
- 2.1 Introduction — 4
  - 2.1i The social/ideological division — 5
  - 2.1ii Unthreading — 6
  - 2.1iii The role of models — 7
- 2.2 Stories of subsistence — 7
  - 2.2i Amboseli National Park, Kenya — 7
  - 2.2ii Owens Valley, California — 10
- 2.3 Economy: divided labour — 12
  - 2.3i The Great Basin, USA — 12
  - 2.3ii Thy, Denmark — 13
  - 2.3iii Neo-Thermal Dalmatia Project — 13
  - 2.3iv Maddle Farm, Berkshire — 14
  - 2.3v South Dorset Ridgeway, Dorset — 15
- 2.4 The Economic Landscape — 16
  - 2.4i The Dorset Cursus, Britain — 16
- 2.5 Socio-economy: states of mind — 18
  - 2.5i Tarragona hinterland, Catalunya — 18
  - 2.5ii North Keos, Cyclades — 20
  - 2.5iii The Southern Argolid, Peloponnese — 21
  - 2.5iv The Biferno Valley, Molise — 23
  - 2.5v Danebury Environs Programme — 25
- 2.6 Discussion — 27

3: Acts of Regional Survey — 30
- 3.1 Historical summary of survey approaches — 30
- 3.2 Determining the land — 30
  - 3.2i How many fingers in the pie? — 33
  - 3.2ii Etching the record: field and form — 33
- 3.3 Survey techniques and methods — 36
  - 3.3i Mapping — 37
  - 3.3ii Survey of the Literature — 37
  - 3.3iii Air Photography — 37
  - 3.3iv Topographical survey — 38
  - 3.3v Upstanding architecture survey — 38
  - 3.3vi Geomorphological survey — 39
    - *3.3via Auger boring* — *39*
    - *3.3vib Exposure recording* — *40*
    - *3.3vic Test pitting* — *40*
    - *3.3vid Machine trenching* — *40*
  - 3.3vii Environmental sampling — 41
  - 3.3viii Geophysical prospection — 41
    - *3.3viiia Magnetic susceptibility and phosphate analysis* — *42*
    - *3.3viiib Magnetic survey* — *43*

	3.3viiic	*Electrical survey*	*46*
	3.3ix	Non-intrusive collection techniques	46
	3.3ixa	*Fieldwalking: generalised regional*	*48*
	3.3ixb	*Fieldwalking: sampling the site*	*52*
	3.3ixc Ploughzone / topsoil sampling		*52*
	3.3x	Intrusive techniques	56
	3.3xa	*Auger boring*	*56*
	3.3xb	*Test pitting*	*57*
	3.3xc	*Trial trenching*	*59*
	3.3xd Excavation		*59*
	3.4	**Spoilt for choice**	**60**

4: Time, space and the thing   61

	4.1	**The perceived**	**61**
	4.1i	Lithics	64
	4.1ii	Ceramics	65
	4.1iii	Architectural material	65
	4.1iv	Environmental data	66
	4.2	**The meaning of "contemporary"**	**66**
	4.3	**What do distributional data tell us?**	**70**
	4.3i	Error and illusion	70
	4.3ii	The meaning of nothing	73
	4.3iia	*Soil and distribution*	*73*
	4.3iii	Quantity and quality	75
	4.3iiia	*East Hampshire, Britain*	*75*
	4.3iiib	*Maddle Farm, Berkshire*	*77*
	4.3iiic	*Tarragona, Catalunya*	*79*
	4.3iiid	*Thy, Jutland*	*82*
	4.3iv	The role of statistics in regional archaeology	84
	4.4	**Find: Fit: Fable — The limits of behaviour**	**84**
	4.4i	Perceptions of constraint	85
	4.4ii	Concrete thought: ritual and economy	86
	4.4iia	*Sergemes Valley, Northern Tunisia*	*87*
	4.4iib	*The Fenland Project, East Anglia*	*88*
	4.4iic	*The Biferno Valley, Molise*	*92*
	4.4iid	*Danebury Environs Programme*	*96*
	4.4iie	*South Cadbury Environs Project*	*96*
	4.4iii	Seeing believing: control and accountability	101
	4.4iiia	*Social authority/ideological authority*	*101*

5: Babel   103

	5.1	**The problem of comparability**	**103**
	5.2	**Categorization of regional research surveys**	**103**
	5.2i	Resource distribution classification	104
	5.2ii	Spatial resolution classification	106
	5.2iii	Chronological resolution classification	108
	5.3	Categorisation and conclusions	109

Bibliography   112

Index   117

1: Regional Philosophies

1.1 Introduction

The goal of this book is not to present a history of archaeological survey, and some will object to the minimal amount of space devoted to its emergence from various traditions. Rather it is a pragmatic attempt to understand what is to be gained from recent work of a regional scale for the purpose of construing the changing patterns of human existence. In the first place I have taken a broad view of what may constitute survey, but have then focused on research projects which have specified their intentions to create at least a description, if not an explanation, of past human activity. I acknowledge fully that the subsequent narratives are contingent products of their time, but within that span a number of traditions are present at various stages of development.

I have chosen to represent three over-arching theoretical branches of regional survey, the first of which is derived from what its proponents would regard as universal principals of a scientific archaeology, whether applied to excavation or a region. Explicitly empiricist, it may be subdivided under such headings as "positivist", "behavioural", "processual", "evolutionary" and "analytical"; it treats landform and human "cultures" as objects of science. The champions of these schools include Michael Schiffer, Louis Binford, Robert Dunnell, David L. Clarke and many others, although these authors are not themselves regular practitioners of regional survey. The second branch relies principally, and unashamedly, on empirical data, but places the human population in the context of a living landscape. This approach, usually derived from the second generation French Annales-school historian, Fernand Braudel, has tended to attract people from geography and history, as well as archaeology departments. Archaeological luminaries include Graeme Barker, John Lloyd and John Bintliff. The third branch also has French roots, growing from the structuralism which so influenced the philosophical thought of the 1960s, with a cast which included Jacques Derrida and Michel Foucault, and which has grown to include Anthony Giddens as its dissemination into the English language has prospered in the shape of critical theory. The shared perception of archaeologists such as Christopher Tilley and John Barrett is that landscape is situated in the perceiver, the human agent.

The presentational strategy of this thesis may seem a little perverse. Rather than creating a story which works through a sequence from beginning to end, I have started, in chapter 2, by looking at the end product, the narrative. At the outset I have proposed a hierarchical set of categories which inform each other as progressively higher levels of narratives are created. After a brief discussion of the narrative classes a selection of actual survey narratives are given in summary form in an attempt to demonstrate the levels of narrative which have been achieved in the past three decades. Chapters 3 and 4 set out to demonstrate how they are achieved. After a discussion of issues which are particular to the general practice of regional survey, chapter 3 lists and discusses techniques, methods and sampling strategies carried out in the field. In chapter 4, the nature of data is considered in the light of difficulties which are peculiar to regional survey, including the problems of dating surface material, visibility and the analytical process. By way of illustration, examples of that process are presented from several surveys, working from artefactual, through primary inferential levels, to higher levels of interpretation. In chapter 5, I have set out to establish a means for comparing the structure, techniques and sampling methods, and the narrative achievements of different surveys.

1.2 Terms of reference

Over the past thirty years there has been an increasing emphasis on regional survey in archaeology. The reasons vary from attempts to gather information from rapidly eroding landscapes, to preventing the destruction of potentially well preserved deposits. There has been debate over whether particular surveys are intensive or extensive - and there has been a wealth of disagreement over what kind of narratives may be gained from survey.

The consequence has been a burgeoning of contradictory accounts with no structured means for evaluating their relative merits. What I have set out to achieve is the provision of a yardstick so that methods can be measured against specific criteria, and so that statistically or inferentially derived narratives can be assessed.

The intention is not to dismiss any specific mode of survey, rather to establish the parameters within which it can pose and answer questions legitimately. In every case a survey is limited by the resources available to it, and but so long as it can present data coherently and make assertions which correspond with those data it has some value.

One problem for anyone approaching the literature lies with the varying choices of terminology; part of my project will be to construct a uniformity of reference. Although authors state that the concrete conditions of each region will require a tailor-made methodology, there is no reason why a common language cannot be constructed and adopted. For this to be achieved we must first consider what might be the object of archaeology.

A scientific discipline is particularised and becomes discrete at a stage when an underlying project or essential spirit of enquiry can be seen to characterise it. The divisions of the natural sciences into the now general pursuits of physics and biology have at their cores elemental concerns: the dynamics of fundamental particles and the origin of the universe; or the origin of life and the medium of replication. At each step, or new boundary, parameters are set within which research is carried out and a fuller account given. Histories of science have shown repeatedly that these parameters exist within ideological settings (Hegel, 1977, 1956; Kuhn, 1970; Trigger, 1989; Habermas, 1978), and that our own perspectives of them are contingent upon the individual interpreter's responses to such purviews.

In chapter 2 I have set out a scheme for classifying narratives according to the level of information they carry. The system cannot claim to be independent of any ideological perspective, although I would claim that it can be applied to any account derived from the widely differing approaches, which I shall summarise here.

Historical studies have shifted from the extremely particular, represented by "great" individuals and events, to an approach from the general. The long term perspective which has informed much of the most systematic regional archaeological survey of the past two decades receives theoretical underpinning from Braudel, placed in the context of earlier writers and refined by more recent writers within the Annales tradition (Bintliff, 1991). What emerges is the convenience of using the *longue durée* as a theoretical framework for those temporal spans lacking contemporary documentation, and leaning towards the medium term for historical phases where guiding world views or moralities are discernible through primary and secondary texts.

Criticism of Braudel's most cited work (Braudel, 1973) has often focused upon his failure to give equal measure to the long, medium and short term. Ironically, it is his failure to deal with the latter two in addressing a period for which contemporary texts are plentiful, which is regarded as a sin of omission. This sense of something absent is derived from a view of the proper role of a discipline, in this case history. Braudel was seeking to redress an imbalance: the unexpressed place of the masses and the landscapes they inhabit. The absence is uttered most sharply, explicitly and implicitly, by those within the Annales tradition. From a very different perspective, spatially-oriented geographers informed by critical theory have complained that it renders "space relatively unproblematical and merely supplemental" to the diachronic dimension, failing to stress the co-situatedness of the "historical" subject/agent and landscape (Soja 1996,172; he quotes with approval Beverly Pitman's phrase "spatialization by adjacency").

The spatial, or horizontal, dimension is emphasised in Bourdieu's concept of habitus (Gregory 1994; 406-410), which informs a perception of full-bodied agents responding to the imperatives of social reproduction in symbolically realised landscapes (Barrett et al. 1991). The landscape becomes inseparable from group and individual identity, although it is perpetually reconceived so that the idea of it may continue to exist. The notion reflects the movement of the Hegelian dialectic where the recognition of self in the other simultaneously transforms, and is, the self. Habitus appears to include Hegel's Spirit and Labour. This may not help us in representing archaeological histories. Whilst we may recognise that there is a subjective dynamic, the discovery of its character can be no more real than the discovery of objective reality. Nonetheless, should we regard it as preferable to the scant flesh on the statistical bones of a regional survey which offers a basic picture of population distribution and means of subsistence (Shennan 1985)?

This issue is not straight forward. In an age where practical, vocational training is emphasised we must consider whether or not there is justification for the practice of archaeology. Whereas even the most rarefied theorising in the natural sciences may lead ultimately to technological gains for people in general, archaeology can offer nothing so concrete. This lack is at its most pointed when we seek funding for research or, indeed, rescue work. What is to be gained for the public at large from an understanding of the day to day lives of our ancestors? Perhaps identity; or employment in the heritage tourism industry; or deeper understanding of our (damaging?) relationship with our environment? Does archaeology have a right of access to funding for esoteric research in the face of an uninterested taxpayer, or a community blighted by unemployment or low wages? If it does it is surely only because of the narratives it can provide, which is bound to implicate the discipline in a dilemma over the purity of science and distortion for public consumption.

Martin Carver has argued (Carver 1996, and elsewhere) that archaeologists need to develop an information-biased view of what has worth within the discipline. His tenet would move us away from the overemphasis on the visible appeal of the upstanding monument, towards processes of understanding based on work in the field, particularly excavation. While a shift towards data collection ought to be welcomed, if the aim is to offer the Public a fuller narrative archaeology, of peopled places in time, it will fail if the excavation of "sites" continues to reinforce existing perceptions of our predecessors as agents within discrete, often very limited spaces.

Many archaeologists reject (or have failed to notice!) the various interactionist perspectives and would regard Carver's approach as too low-key. Andrew Sherratt, has argued that the local or regional picture can only truly be understood at a macro-regional level, set on the scale of the grand narratives which shaped archaeological thought for much of the first half of the 20th century. He emphasises the spatial dimension against a model of development in stages, asserting that the opportunities of location may bring about transformation whereby for instance, an unproductive backwater may eventually outgrow two larger cultural centres by developing items with added value which may seduce passing traders (Sherratt 1995, 17-18). He stresses that it is often the exchange of exotica which generates change. Defining Homo Sapiens as a creature of "wants" rather than mere needs, he argues that "without wants, the [socio-economic] structures are merely latent" (Sherratt 1995, 22). If Sherratt's desire for the grand narrative are to met there are implications for regional survey. Macro-regions cannot be surveyed within single field-based projects, but must rely for their narratives on reviews of the literature. This, in turn draws attention to the issue of comparability of field techniques, sampling and analytical units.

Sherratt's mentor, David Clarke, was an archaeologist who wanted to tell well-founded stories. He identified that foundation in the artefact as the fundamental unit of research. To create a story we must be able to gauge the potential of the artefact, to give it a biography with which we can build a picture of the practical limits to production which could have been achieved at a given time, in a particular place. But it is easier to consider an object's biography at the level of utility, rather than its significance - its symbolic potential. If we are to understand how people conceived of themselves, we need to consider the fullest meaning *for them* of their artefacts.

In creating these biographies it is difficult to avoid resorting to analogy, which will show us what the object and its user could do, but not necessarily what they did do! Ethnography has been the favoured informant of analogous reasoning for practitioners as diverse as structuralists and post-structuralists (Levi-Strauss 1963; Hodder 1982), through processualists and behaviouralists (Binford 1978; Schiffer 1978; Schiffer et al. 1981) to spatio-analyticists (Foley 1981; Bettinger 1977)! The common hope is that ethnographic data may generate models of activity associated with particular artefacts and inter and intra settlement patterns which may be tested by appropriate field strategies. In using such models we must take care to isolate those aspects of productive behaviour which generate durable artefacts and note the means by which they enter the record. In some cases projects have had access to recent ethnographic literature relating to their regions (Bettinger 1977), or may have had the opportunity to observe contemporary populations using technology similar to that available to their predecessors in the particular region, although not necessarily a similar landscape (Foley 1981).

Whatever the approach of the authors, all archaeological narratives must ultimately be rooted in work on the ground. In chapter 3 I have described a range of techniques, various combinations of which form the basis for fieldwork in modern surveys. The list is not comprehensive and, in any event, new techniques or the refinement of old ones will soon extend it.

In chapter 4 I have focused first on data, and on what we may choose to perceive or treat as such, and have gone on to look at how different practitioners have selected it and used it to create regional narratives. Inevitably there is discussion of the temporal and spatial dimensions, and of the meaning of "distortions" of the data, whether due to their mode of distribution, or to the skills of the fieldworkers.

The final objective of this book is to present a means for comparing and assessing different surveys. To this end I have classified projects according to resource distribution, their intensity of fieldwork and detail in recording and mapping, and the tightness of their chronologies. These categories by no means necessarily reflect the quality of survey work. Spatial and temporal resolutions frequently vary according to the conditions of the study area, and the period targeted.

The last chapter presents as a case study an actual project design; that of the South Cadbury Environs Project. The chapter summarises the reasons for choosing that region, and outlines narrative objectives. The survey's detailed exploratory fieldwork, some of which is presented in the pertinent sections of other chapters, informs the explicit proposals made for future work. The chapter is not aiming to say "this is how it should be done"; rather, the intention is to raise issues which should be considered when dealing with a variety of different landforms which present diverse challenges and potentials for fieldwork.

2: Landscape Narratives

2.1 Introduction

Karl Popper has referred to disciplines such as sociology and psychoanalysis as "pseudo-science" (Popper 1980) and nowhere is archaeology's vulnerability to such a charge more apparent than in the accounts we give from the most indirect of experience. Certain orders of narrative do prove to be falsifiable, though not always in a manner which an author might have expected. When Clarke (Clarke 1972) chose Bulleid and St. George Gray's ambiguous records of Glastonbury Lake Village to construct a paragon of spatial interpretation he must have anticipated debate from within, or close to, his theoretical paradigm; but his argument was unable to contend with old fashioned reviews of the evidence put forward by Tratman (1970) and Coles (Coles & Coles 1986).

Most archaeologists will accept as *a priori* the dismissal of Arthurian mythic excesses, but are they right to be more accepting of the fashionably psychologised landscapes of Barrett's Neolithic Wiltshire (1994) Tilley's South West Wales and Wessex (1994) and many others. Should we not prefer the responsible adherence to the empirical of Shennan's (1985) survey of East Hampshire? Before attempting to address such questions we should be careful to distinguish between the evidence of the material record and the material means for subsistence. A distinctive form, a form with sufficient identity to be reproduced, must have been conceptualised; once it has existence as a concept, as Duchamp's toilet has shown most effectively, it may enter new areas of discourse and acquire a significance quite other than any borne of physical need.

For many authors, archaeological story telling is hierarchical. Raw data inform, even if only rarely can they be said to necessitate, an interpretation which, placed alongside other interpretations, forms the platform for the next layer of analysis. Thus, from surface collection a given number of artefacts will be assigned the label site; other sites matching the criteria come to indicate settlement patterns; population growth is assessed, and with it the efficacy of a particular productive mode within the constraints of a landscape, inferred from environmental data. Thereafter settlement/ activity distributions may be clothed in economics, politics and ideology.

Despite the unsettling lack of stability inherent in the structure, archaeology's legitimacy is based on the assumption that it *reconstrues* the past, and does so inductively. Yet despite the malleability of the data which can so easily protect us from accountability, the paradigmatic frameworks within which we gather and process information constantly change, as indeed they do in other areas of science. A major cultural shift within archaeology is that from research concerned with a particular feature (site), to the landscape - sometimes associated with a feature, sometimes chosen as an arbitrary example of a particular geological/ ecological range. Within this shift are regional studies borne of empiricist traditions, social and spatial archaeology, the Braudelian physical and Bourdieu-derived mental landscape archaeologies. While many practitioners are avowedly eclectic, blood has been spilt amongst those of polemical tendencies. Arguments range from wrangles over whether surveys should be extensive attempts to collect sites or intensive surface collections from which regional inferences can be drawn (Hope Simpson, Cherry in Keller & Rupp 1983), to the assertion that there is no such thing as a site other than as a piece of land subjected to a particular form of research.

What follows is a presentation of narratives resulting from various forms of area survey. My prime concern is to address accounts which have few or no primary texts to test their arguments, although because Mediterranean land has been surveyed so frequently there are several cases where the authors have been in a position to draw upon Classical textual sources. My analysis will be layered according to the remoteness of an explanation/interpretation from raw data (table 2.1).

Table 2.1 - Classified narrative sequence

	Social	*Ideological*
Class G	Political (G1)	Belief system (G2)
Class F	Economic (F1)	Ritual practice (F2)
Class E	Distribution of behaviour	
Class D	Artefact determined behaviour	
Class C	Artefact chronology	
Class B	Artefact / Ecofact type	
Class A	Artefact / Ecofact	

Each class is characterised by particular theoretical attributes which reflect its place in the conception of narrative. Each has a meta-theoretical, *a priori*, assumption from which specific theories are formed.

Class A meta-theory assumes that an artefact is a definite material entity which has been modified by human behaviour. Thus most baked clay objects are artefacts, as are many chipped stones, all processed metals, and so on. Many ecofacts fall almost as easily under the same heading. Carbonised grains are usually the result of human agency, typically drying processes. Less obviously, many organically rich soils are artefacts of manuring and other cultivation practices.

Class B meta-theory assumes artefacts may be grouped as *types* according to their common characteristics so an assemblage of chipped stones may be identified because of their proportional length against breadth. A subset of the total assemblage might show long, slightly concave, and conchoidal scars. At Class D level these might be called respectively flakes or blades, or cores.

Class C meta-theory assumes that the form, fabric and embellishment of an artefact is constrained by the time and place of its manufacture, and by the cognitive world of the agent making it. In general stratified deposits are presumed to represent the form, fabric and decorative preferences of artefacts in the approximate sequence of their discarding. It follows that artefacts from unsealed contexts may form the basis for meaningful maps of behaviour distribution when comparison with artefacts from sealed deposits allows inferences to be made about their date of manufacture.

Increasingly absolute dating techniques (radiocarbon, dendro, thermoluminescence) allow relative artefact sequences to be tied to a specified chronology.

Class D meta-theory might appear to stand logic on its head by suggesting artefact use is not the direct consequence of an agent's planned intentions. Of course the maker's and the user's intentions are part of the story, but the archaeologist finder needs to be aware how the form and fabric of an artefact constrains its perceived utility, from the moment of its manufacture to the moment of its discard. Such constraints upon perceived utility are both absolute (physically determined), moral (determined by social praxis), and contingent on time and place.

An archaeologist's insight into the perceived utility of an object at the time of its use may be gained analogously through everyday experience, through experimentation by reconstruction, through ethnographic observation, and by reference to records of the above and of other archaeologists' insights, whilst recognising that all such insights are contingent on the time and place of observation. A simple example might be a theory that a baked clay artefact which partially encloses space has the potential to store, and to be the dispensary of, other materials in a manner which may be categorised by analogy with modern forms (e.g., "bowl", "dish" etc.). At a different level theory would acknowledge that an artefact's associations with other objects and events, both artefactual, organisational and naturally occurring, during the period from its manufacture to its discarding, will effect contemporary perception of its utility.

Class E meta-theory assumes that artefact discard behaviour is determined by physical and moral potentials and constraints. The present spatial distribution of artefacts is determined by the locus and manner of an agent's behaviour at the moment of discard, by the physical properties of the artefact and the material around it, and by subsequent episodes of soil movement due to natural, faunal and human activity, including that at the moment of recovery. *Intuitive* and statistical inferences may be made about the significance of their relative distributions by mapping the present location of the various categories of artefact.

A further theoretical assumption is that a particular range of artefacts perceived to be associated in time and space will provide insight into some of the potential for behaviour available to the discarders of those artefacts. In this way it may be possible to diagnose areas reserved for the disposal of waste, domestic activity, tool production, hunting, agriculture and ritual practice.

Class F1 meta-theory assumes that from analyses of class C, D and E hypotheses, in conjunction with ethnographic observations and modern socio-economic theory, it is possible to make inferences about past modes of physical and social production, subsistence and exchange. A simply stated hypothesis within this framework might be that incidence of poor meat bones at one settlement, and good meat bones from the same species at a neighbouring contemporary settlement indicates that the former is producing food for the latter. Another example might be that a high incidence of high investment artefacts (finewares, imported pottery, gold, stone axes made from non-indigenous material) implies that an place was used for activities carried out by individuals or groups of high status.

Class F2 meta-theory assumes that from analyses of class C, D and E hypotheses, in conjunction with ethnographic observations and landscape artefacts, it may be possible to make inferences about modes of past ritual behaviour. Ritual practices may leave material traces, either in the form of particular, specialised artefacts, or of special configurations of mundane objects.

Class G1 meta-theory assumes that all social systems require the making and communicating of decisions by either the whole group, a section of the group, or an individual, and that there will be rules governing the succession of decision-making. It may follow that if decision-making is consensual there will be few or no signs of distinction between members of the group, whereas if decision-making is by a minority of the group or an individual there are likely to be material signs of differentiation of status.

Class G2 meta-theory assumes that many social systems incorporate systems of belief which may be explanatory and/or moral. It may follow that consideration of the loci of monuments or deposits, including the particular natural and anthropogenic characteristics of the space and of the rituals interpreted to have been associated with them, may allow inferences about belief systems.

This scheme does not pertain specifically to regional survey, but to archaeology more generally. The basis of narrative structure should not necessarily differ, whether pertaining to a single excavation or to a region, although certain narrative classes, notably D, have an especially prominent role in regional survey. Frequently, in practice, there has been a different treatment of artefactual data, with much greater investment of resources in the analysis of excavated data rather than in that derived from surface collection. As should emerge in the later chapters, in particular chapter 4, I regard this as an unwarranted defect in analytical practice.

2.Ii   *The social/ideological division*

The division of classes under *social* and *ideological*, while not entirely arbitrary, is by no means clear cut. Social refers to those classes of indirectly perceived activity which are generally held to take place, and to be accountable, within the living human sphere; they structure authority and determine the character of agency for individuals and groups.

Ideological here refers to those classes of indirectly perceived activity revealing phenomena which invest with authority all spheres of life, including the political and economic. The source of its authority may sometimes reside in what is perceived as its derivation from beyond the human living - for instance, through tradition or religion. The word is used here to indicate those moments of behaviour and individual or group self-perception where the authoritative rationale is made explicit. Thus, when an English monarch addresses the House of Lords he/she is acting within the social sphere; but a public view that he/she acts within a 'god-given' role to invest parliament with his/her authority is ideological. This, as Barrett would point out, "is not a 'false' consciousness' but a dominant discursive reading of key cultural values" (Barrett et al. 1991, 7).

In the field an archaeologist will frequently make experience-based judgements concerning classes B to D - and less frequently F, as well. In short, the narrative structure includes feedback loops so that, for instance, class D information obtained earlier and, frequently, elsewhere may inform class B inferences in the field. This compression of explicit narrative structure is almost inevitable yet, as will be shown below, it can have dramatic consequences for the day to day operation of a project when, for instance, a scatter of variously dated pottery within a limited area may be used to diagnose a settlement site.

Even class A should be seen as integrated into the narrative process, not simply precipitating it. An object may not be collected or counted (i.e. regarded as an artefact) because either an inexperienced fieldworker fails to recognise it as an unconformity, or an experienced fieldworker regards it as a class of data without value to the particular project. In the worst cases the material will be collected, but then discarded without record.

2.Iii Unthreading

While few archaeologists possess the art of the seamless text, those surveying regions have come to view the disjunctions, the discrete cultural and temporal entities which still inhabit the world of the excavation report, as a poor reflection of reality. The current trend towards emphasis of continuity within a temporally, as well as a spatially, defined landscape (Bintliff 1991; Foley 1981, 29), has generated field and analytical methods which afford greater explanatory power within interpretation. In some respects this represents a shift towards an older, almost antiquarian, tradition where the desire for narrative meaning is at least as strong as the desire for evidence. Although in England, and many other countries, this trend is fuelled by the heritage industry, often with a prod from government, the impatience with bare statistical outlines is growing amongst archaeologists: "... while we applaud the fact that the character of the surface archaeological record can today be described more accurately than a decade ago, it is time to devote more time to explanation" (Cherry et al. 1991, 327).

Recent publications of regional survey show a willingness to recognise political structure, organising and controlling a social group and establishing a distinct identity from others. In the Biferno Valley, Barker suggests a progression from inter group support, reinforced by prestige gifts, in the Early Neolithic, to ranked societies where the highest placed agent controls "aspects of the production, distribution and/or consumption of resources" by the Mid to Late Neolithic. (Barker 1995, 127-129). Jameson et al. (1994, 341-45), with the Southern Argolid as their subject, while recognising "imported" raw materials from as early as the 7th millennium BC, and suggesting that crops susceptible to dry storage might constitute trading commodities, resist the temptation to infer any particular social structure. The quality of evidence from the comparable periods covered by the two surveys is little different; but one group of authors displays a greater willingness to treat complex layers of interpretative analysis as *data* from which a new narrative layer can be composed. The other group is more reticent!

The application of common sense, of analogies derived from one's own or another's experiences, allows an archaeologist to build a repetoire of recognisable forms - artefact, structure, frequency, distribution - all of which are important in guiding collection and analysis of data, yet any of which may prove inappropriate, or which may not be disprovable. A salutary distinction of one interpretative layer from another might be based its paradigmatic durability. Class B paradigms have a long shelf life, although particular examples may prove more resistant than others; perforated lumps of baked clay from British Iron Age contexts have long been designated "loomweights", and continue to be so; but there is an incipient trend to classify some at least as "oven bricks" (Cynthia Poole in Cunliffe 1995, 285-86). For many there are only a few means for discriminating between flint artefacts and flakes, and cursory attention may be given to whether or not the piece is a scraper and part of a concentration, so indicating a domestic site, or an adze which may have been discarded further from home.

Class C paradigms are routinely treated to internal adjustments, but rarely to a major overhaul. The availability of ^{14}C dating has made considerable impact on some typologies, and dendrochronology's effect may prove even more dramatic; but the all important sequential phases are comparatively unaffected. Indeed, the existence of these scientific resources can induce the archaeologist to hand over responsibility for reconsidering chronologies to technologists from outside the field. Nonetheless, absolute dating has an increasing role to play in our judgement of what might have been contemporary, a crucial prerequisite for well-founded hypotheses about human interaction of particular relevance to regional survey.

Class D paradigms are much less stable. Beyond the immediate response to the appearance of an object, the experimental archaeologist seeks to show either what it can or what it cannot do; our behaviour is determined by the object's functional potential. We all recognise the flint axe - but was it used for cutting trees, as a hunting implement, or as a prestige gift; a ceramic sherd was part of a vessel, but was it for storage, carriage or cooking? In some cases the paradigms are not clearly established; there is no firm view with which to agree or disagree. Class E reflects a compounding of this problem; not only do we suffer the ambiguities of potential - we must contend with the problems of how widely that potential can be exercised, and to what degree the distribution of artefacts is a function of their potential as combined entities. In the natural sciences a paradigm fails at the point where it can no longer accommodate a new range of data. The character of archaeological fact is often too ambiguous to present an effective challenge to existing paradigms.

The greater resilience of classes F and G has much to do with the manner in which we structure our perception of the present and are infected by it. It is hard to resist the most successful dynamic explanatory models, often with a strong tradition dating to the 19th century. Ultimately, these classes are likely to be shaped by the individual researcher's particular *weltanschaung*; more interesting is how that purview influences the analyses which generate classes D and E - in other words, how the "final account" shaped the relationship between the paradigmatic classes B/C and the less structured D/E.

2.1iii   The role of models

For the purposes of later analysis it is appropriate to clarify levels of models as I propose to define them. The explicit framework of discourse chosen is the sub-paradigm, where "archaeology" is a discipline in the Kuhnian sense. Under that classification are *processual, behavioural, spatial, problem-solving, evolutionary, social reproduction, structuralist/landscape*, etc., archaeologies. At this sub-paradigmatic level I wish to distinguish two general categories: those which have a direct relationship with field procedures, and those which are principally means for structuring interpretations, although they may inform the researcher's perceptions in the field. The former category exemplifies a hypothetico-deductive approach, whereas the latter tends towards inductive reasoning. It is not possible to classify these paradigms within stricter scientific traditions because any may draw from more than one such tradition, or may eschew scientific modes of narrative altogether.

Within each paradigm will be meta-models - general models which might be applied to a variety of regions and studies within them. Characteristically, the problem-solving paradigm addresses existing narrratives of and within a received view of a region's archaeology. For example, after providing a geological, environmental and historical background Carreté et al. (1995, 9-44) pose a series of questions, summarised below (*2.5i*). The spatial paradigm holds that the distribution of artefacts is continuous, but of varying density. Where taphonomic processes have not intervened to a significant degree, Spatial Components are meta-models of an explanatory kind, from which interpretations of behaviour may be inferred. These meta-models exist as the parameters within which case-specific models can be constructed. Good examples of the latter are Foley's outlines of Maasai Pastoralism, Pastoralist Neolithic and Hunter-gatherers (see *2.2i* below).

John Bintliff (1997, 1-38) has investigated the aptness of specific meta-models, applied in the Aegean, which are designed to function at a regional level. Leaning in the direction of the structurally interpretative, he concludes that Core-Periphery models, where production in a "fast stream" region outstrips surrounding regions by gaining early political, economic or ecological advantages are useful in specific cases, but that in general eco-demographic models are more powerful tools.

It is only at the level of the case-specific model that the archaeologist can form hypotheses about the nature of residues (their qualities, quantities and distributions) he/she anticipates from fieldwork or reviews of the literature. I shall be arguing that few, if any, archaeologists approach projects without a model/hypothesis in mind, but that it is a recent, and growing, trend to acknowledge, explicitly, and to systematise what are, in effect, prejudices informing perception. Following Gadamer (1975) and Ricoeur (1981) I shall argue that prejudice is a necessary foundation for the structuring of understanding.

Case-specific models are not necessarily constructed in anticipation of a nearly identical narrative. In the Amboseli, Foley is able to reject two models, and has to accommodate divergences from the artefact distributions predicted by his third model. Chapman et al. make this explicit when outlining what may be expected of a Land Use Capability (LUC)-type model:

> "This model makes uniformitarian assumptions about the economic strategies underpinning land use; to the extent that these assumptions are false, the actual data may well deviate from the predictions" (Chapman et al. 1996, 253).

In the following sections (*2.2, 2.3* and *2.4*) selected narratives have been outlined with reference to their paradigmatic perspectives. Preliminary remarks are made concerning their structures, but the principal function of these sections is to illustrate the range of narrative types to be found in the evolving literature of regional survey. A much fuller critical analysis is to be found in chapter 4, relating the application of models to data, and to sampling procedures.

2.2   Stories of subsistence

2.2i   Amboseli National Park, Kenya

Robert Foley's work in Kenya's Amboseli Park remains a beacon in regional survey. His report (Foley 1981) of the theory, method (in field and in subsequent analysis), and modelling is exemplary in its detail and clarity of exposition. The project was one of the first to take into full account the geomorphology and ecology of the research area. Careful consideration was given to the length of time sediments were open to discarded artefacts (Foley 1981, 182), and to whether they presented depleted or aggraded surfaces, and the implications for the incidence rate of surface artefacts. Through use of existing data Foley was able to model the chronologically variable "productivity of the landscape, due to factors of climate, topography, soil and vegetation in relation to biological and social requirements" (Foley 1981, 12).

Ethnographic data, in particular from observations of the local Maasai, allowed the construction of models from which predictions could be made about discard patterns: thus the artefact, "any material that has been modified by human activity, in terms of size, shape and location", as "the basic independent unit of archaeology" (Foley 1981, 11; Clarke 1968, 134), is the appropriate resource with which to implement testing. Table 2.2 outlines three models of behaviour and consequent discard distribution which guided his survey methodology and structured his narrative. The varying degrees of population group mobility explicit in each model required a "regional off-site" approach for "ephemeral but numerous open sites" (Foley 1981, 8).

The absolute criterion for spatial model suitability resides in artefact distribution. Table 2.2 shows that the Maasai pastoralist model anticipates that, allowing for a sloping

Mode		Maasai patoralism	Pastoralist Neolithic	Hunter-gather
Settlement distribution	S.C. - A	Preference for basement slopes in north and along south edge; non re-use of specific sites	Similar to Maasai but more constraint due to need for water, particularly when domesticates first arrived	Kittura Hill (for viewing); swamp at Longenya, with access to south woodland; north slopes in wet season
Settlement density	Frequency	13 per annum in north; 12 per annum in south		? 1 permanent band, 3 dry season of approximately 25 people, moving on approximately every 4 weeks (Hazda transhumance)
	Size	3022 sqm	450 sqm - 25,000 sqm	
Settlement rate	Duration	3.7 years in north; 5.2 years in south	Similar to Maasai, but parasite populations may increase rate of abandonment	4 weeks
	Increment	3.51 per annum in north; 2.31 per annum in south	Higher dependence on and exhausting of local wild resources may increase movement rate	3 x 6 and 1 x 12 = ca. 30 new settlements per annum (within Amboseli ca. 200 sqkm suited to pattern
Time span		400 years	>2000 years	Unknown
Residual density		Total: 2388 sites. 5.44 per sqkm.		In 1000 years, 30,000 settlements: rate of 150 per sqkm per 1000 years.
Settlement periphery	S.C. - B	Minimum peripheral discard = settlement area x 2 x 2388 sqm	Similar activities to Maasai, so proportional to that figure	Short duration will reduce volume of discard
Secondary homerange foci	S.C. - D	Meat-eating locations, usually shaded, grazing next to swamp, woodland and waterholes. Low discard rate	Increase in dependence on hunting, high and dispersed rate of discard, particularly where high secondary biomass	Direct relationship with food resources of swamp and south woodland, blending with settlement zone to define "major areas of exploitation". Moderate discard around water supplies
Occasional loci	S.C. - C	Good grazing land and close to settlement. Similar to S.C.-D	Similar to Maasai, depending on proportions of hunting and gathering	Adhere closely to distribution of resources, including less intensively exploited areas
Predicted distribution		High density along north slopes due to settlement and wet season grazing. Secondary distribution in extreme south of area. Low around waterholes and swamp edges. Metal technology so poor artefact survival	Higher along north and south boundaries of study area. Slightly less sharply defined boundaries of S.C. A, B, C and D discard. Latter likely to moderate densities in south woodland and swamp. Generally high density	Highest in swamp region and south west hills. Moderate on south woods and perhaps in woods existed before caliche formation. Very low in central basin and north slopes. Density greater than the Pastoralist Neolithic
post depositional effects		Sloping topography in north. Aggradation in south; but recentness of occupation in mitigation	High density of durable artefacts mitigate poor visibility on north slopes, otherwise visible	Discard expected to have been onto Pink Tuff, so post depositional processes even exvcept where floding occurs

Table 2.2. Three models of subsistence and predicted artefact distributions (Amboseli Park)

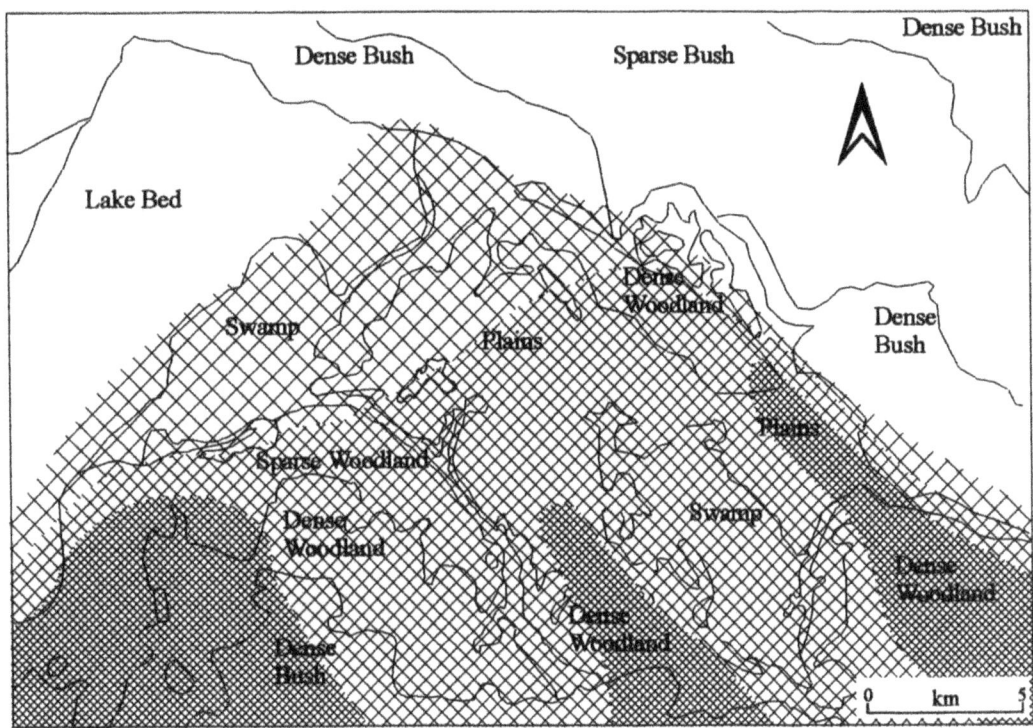

Figure 2.1. Predicted relative artefact distributions superimposed upon the reconstructed ecology of Amboseli Park (after Foley, figs. 4.6 and 6.4).

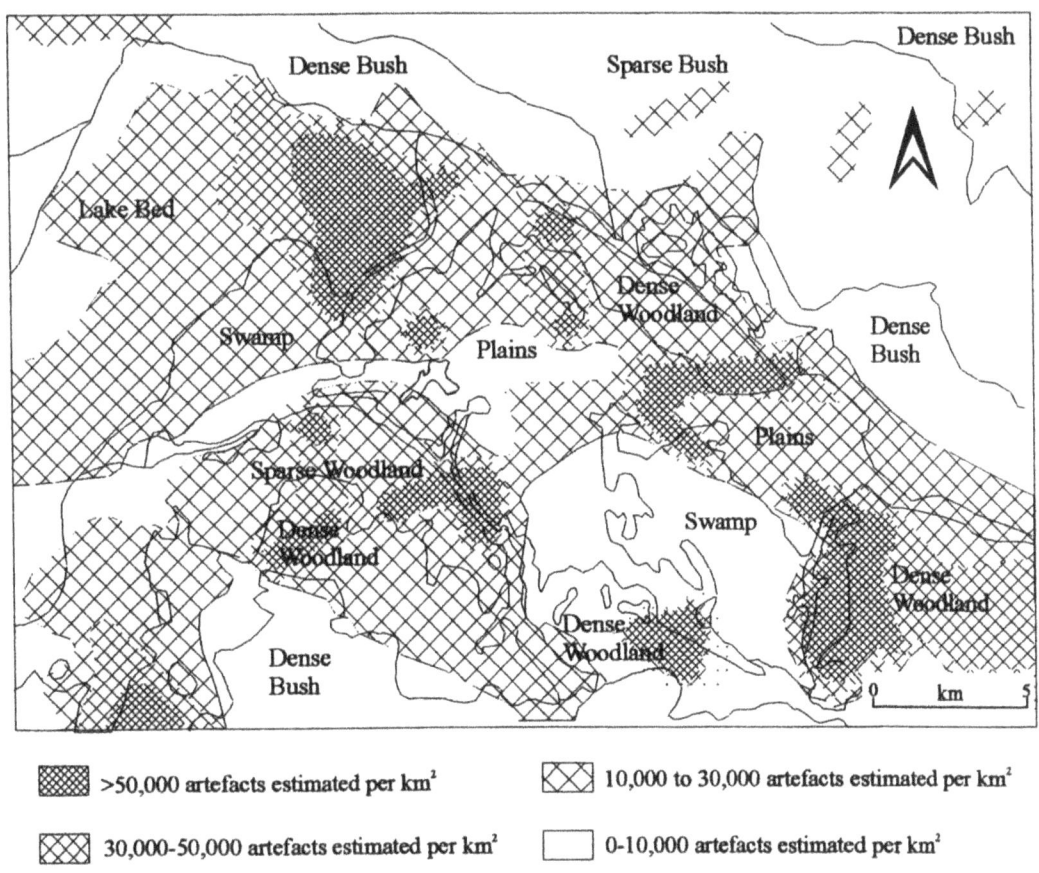

Figure 2.2. Recorded artefact distribution superimposed upon the reconstructed ecology of Amboseli Park (after Foley, figs. 4.6 and 7.3).

northern topography and an aggrading southern geomorphology, the highest densities ought to occur in the settlement and wet grazing areas along the northern slopes, with secondary-type densities in the extreme south. Low densities would occur around water sources, but there would be very little on the central plains and in the eastern part of the survey area (Foley 1981, 110).

Expectations of the East African Pastoralist Neolithic model are broadly similar to those for the Maasai in the highest and middle ranking densities, although hunting should have left a more gradual decline from middle to low density areas. Once again, a population restricted by the need to water stock, would be expected to discard little in the plain or eastern survey area.

The hunter-gatherer model predicts high densities in areas of woodland and swamp, as well as in the south western hills from which there is easy access to those resources. Moderate densities will occur in the southern woodland zone, and further north if it was wooded there too. Artefact incidence ought to be negligible in the basement hills of the north area.

A total of 8531 (lithic) artefacts was recovered from 257 sampling points (Foley 1981, 119), and was computed as densities per km^2, by which it could be shown that "artefacts are virtually continuous but differentially distributed through out the study area" (Foley 1981, 125). Only 100 pottery sherds were found, probably an under-representation since the sampling strategy did not take account of its likely settlement-specific discard location (Foley 1981, 160). Bone, although frequent, and sometimes clearly associated with a particular sediment, was not the subject of detailed analysis. Unfortunately the artefacts offered poor chronological resolution - the very small ceramic sample and broad date range for the lithic assemblage militated against clear phasing. However, one of the great strengths of the field and analytical methods was the calculation of sediment-specific discard rates on the basis of the known duration of formation for some sediments.

Material recovered from a Pink Tuff sediment exceeded the next most prolific, a Brown Tuff, by a factor of nearly five. Indeed, where the Pink Tuff had eroded completely, some Brown Tuff finds were likely to have been discarded while the Pink Tuff was being laid, apparently between 3000BP and 1000BP. Although similar processes indicated that some artefacts from the lake bed were also likely to be broadly contemporary, material from that surface had a much longer temporal span of discard; in any event, there were far fewer finds, reliably indicating low discard rates during the deposition of the Pink Tuff. Thus Foley was able demonstrate a sufficiency of data from which to make fairly reliable statistical inferences about the spatial character of human activity in the area for a particular 2000 year time span.

With data for present and past vegetation, and Western's (1973) information about habitat preferences for specific fauna, Foley was able to show "the distribution of biomass in the Amboseli at the present time with a view to predicting the spatial patterning of human groups that may be exploiting them" (Foley 1981, 66-71). Using a combination of biomass models for the present with ecological models for earlier phases he was able to create more detailed models of discard patterns derived from hypothetical modes of subsistence.

Analysis of Stage 2 experimental fieldwork indicated that, aside from a sample points which were excluded from fieldwalking due to their high vegetation density, only grass cover caused significant reduction in artefact recovery rates. This problem was due to sustained heavy rainfall in 1978, following a seven year dry spell (Foley 1981, 162-63; Table 8.1).

No one model fitted the data, but Foley concluded that hunter-gatherers, probably with a "bow-and-arrow technology" exploited', woodland game, rather than that of the plains. During the dry season up to three competing bands of approximately 25 people "maintained the stability of the settlement pattern" in three "broadly similar habitats" necessarily within reach of water, and close to hills. Only with stresses such as increased soil, and therefore vegetational, salinity, did pastoralism develop, probably as a supplement to hunter-gatherer subsistence (Foley 1981, 91). Figure .1a shows an ecological outline of the study area overlain with the predicted distribution of hunter-gatherer material; Figure 2.2 shows the recorded distribution after survey (after Foley 1981, figs 4.6, 7.3 and 6.4).

Despite the detailed survey work there is no account at class G level of either the social or ideological. A very general economic account is given, but this reflects the very limited class D and E information; Foley acknowledges that further work "on the function of stone artefacts may well enhance the resolution of spatial analysis", but the low incidence of tools throughout the survey area (for instance, a total of 92 scrapers) necessitated reliance upon "gross artefact densities" to gain a statistically viable sample size (Foley 1981, 188).

2.2ii   Owens Valley, California

A broadly similar approach, again using a method based on the stratified sampling of ecological units (or biotics), was employed by Bettinger (Bettinger 1977) in his survey of Owens Valley, California. Situated in the Great Basin, the valley is a trough formed by mountain ranges on its west and east sides, with four main ecological resource zones. Riverine resources included hydrophytic vegetation, and amongst the fauna were mollusca, fish and migrant waterfowl. The lower valley sides comprised desert scrub of low shrubs and seed-bearing grasses, supporting small rodents and a few antelope throughout the year, and deer and mountain sheep in the winter. At above 6500 feet foothills on the east offer open forests of the edible nut-bearing pinyon pine, populated in the summer by large ungulates. At its upper limit the pinyon pine is interspersed with open meadows in depressions in a sagebrush zone which, although offering few edible plants supports deer, mountain sheep and antelope in the summer and early autumn (Bettinger 1977, 3-5).

Ethnographic accounts published over a period of forty years (Steward 1929-70) indicated that "11 annual activities that were fundamental to the settlement subsistence patterns" (Bettinger 1977,9; Table 2), may be treated in terms of spatial distribution. Frequency of occurrence and the range of artefacts they require. It had been argued that this Desert culture of intensive but unspecialized exploitation of food resources, which had persisted through historic times, had altered little since 800 BC, and had been "unaffected by long-term post-glacial climatic fluctuations" (from Jennings 1957, 1964, 1968). This model could be tested against the

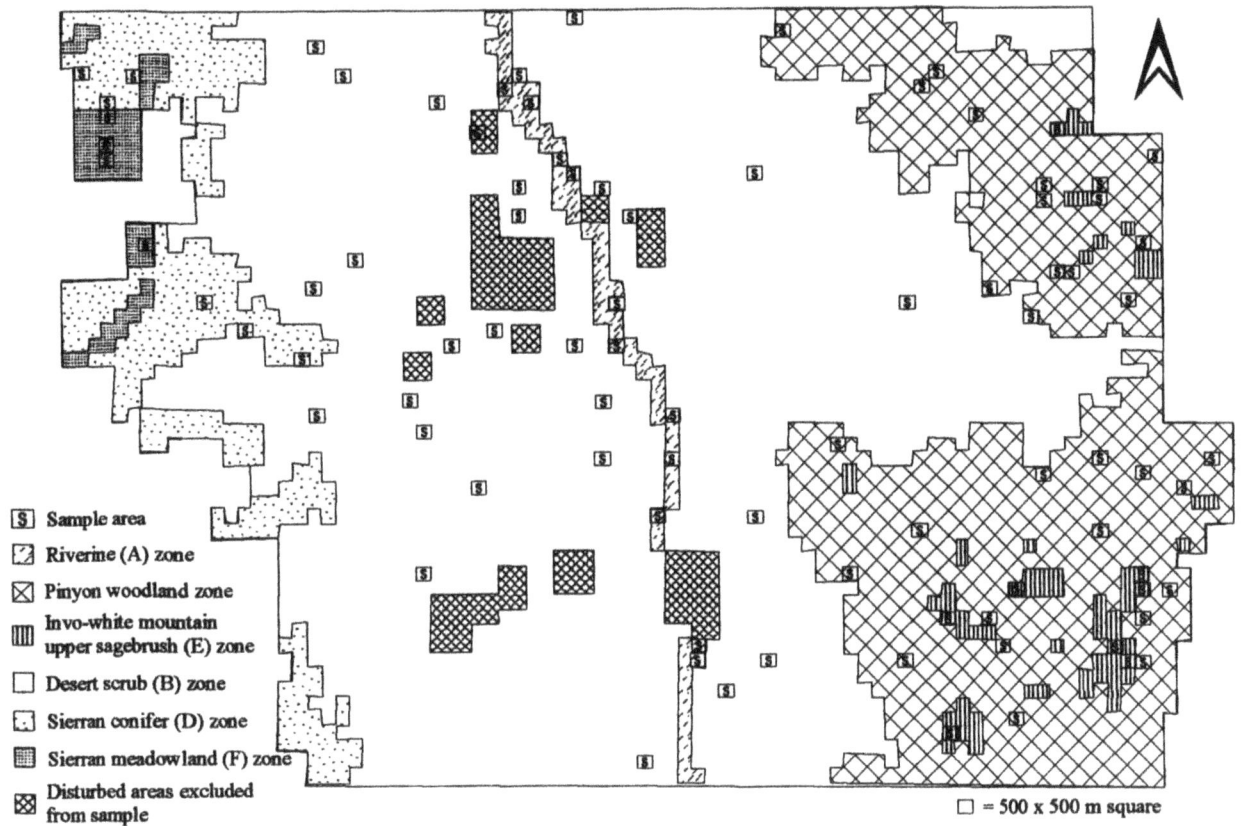

Figure 2.3. Biotic strata with sampling grids at Owen's Valley, California (after Bettinger 1977, fig.3)

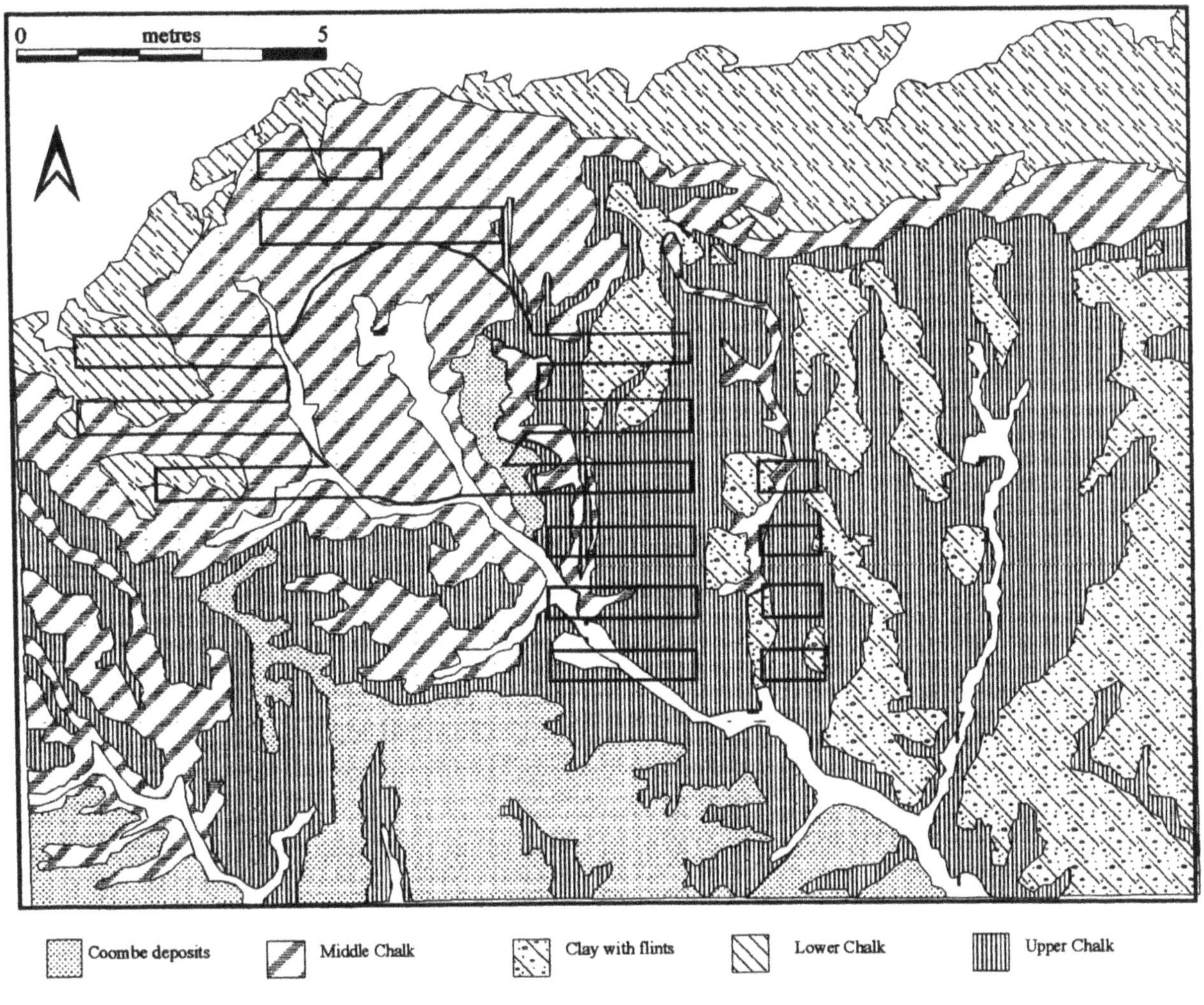

Figure 2.4. Geological map with sampling areas, Maddle Farm, Berkshire (after Gaffney & Tingle 1989, fig. 2.1)

contention (Heizer 1956 etc.) that subsistence varied with ecology, included specialisation, and was susceptible to climatic changes (Bettinger 1977, 3).

Previous archaeological work had demonstrated the relative visibility of nine classes of feature/artefact which, in different combinations, might reflect fairly particular activities. A control was introduced, testing the appropriateness of linking an inferred activity to a specific ecological zone (Bettinger 1977, 12; Table 6).

The artefact assemblage afforded broad band temporal discrimination. Settlement-specific features/artefacts provided evidence for some continuity of preferred occupation zones, but considerable variation of patterns for other activities. During a period from 1500 BC to AD 600 there was some movement of occupation sites from riverine to desert scrub zones - perhaps due to population growth, although climatic evidence indicates higher moisture and, with that, greater productivity from scrub plants. It was only after AD 600 that pinyon nuts were collected routinely, which may again reflect population growth or, equally, reduced productivity from the scrub fauna. Irrigation systems, seemingly constructed after AD 1000, may have diverted communities from hunting, so accounting for the "disuse of upland and desert scrub temporary camps" (Bettinger 1977, 15). On this account the Desert culture model fails to predict both the lack of continuity in the activity pattern, and change precipitated by climatic conditions.

In the above examples the landscape has combined with the climate to provide ecological zones which, although not chronostatic, have remained within a recognisable pattern. In regions where human activity has persistently modified the landscape environmental controls, against which artefact distribution can be measured, are more variable. Where long sequences of environmental data exist the consequent narrative may be more detailed (Coles & Coles 1986); however, in much of Europe, for instance, the varied details of subsistence, and their very local specificity, renders the forming of truly regional ecological narratives problematical. Equally, surveys are often over extensive areas of aggrading lowland (Hayes & Lane 1992; Lane 1993) or on badly eroded upland.

2.3 Economy: divided labour

2.3i *The Great Basin, USA*

The survival of the almost mechanistic approach to method and the subsequent narrative is documented in a recent review of the archaeological literature counterpointed by a highly chronologically resolved climatic study. This application of finer climatic data, derived from tree-ring analysis, has been used within a problem-solving technique to address the "boom or bust" characterisation of human subsistence in the Black Mesa and Vermillion Cliffs regions of the northern part of the North American Southwest (Larson et al. 1996, 217).

Where Foley espoused the ideology of spatial archaeology, which has a direct link with empirical field methodology, Larson et al. carry forward the banner of evolutionary archaeology. In this the latter they are closer to John Barrett, since the perception of evolution, like that of reproduction, is biased towards a narrative rather than a methodological structure; reproduction and evolution are inherently explanatory concepts, whereas spatial is descriptive. On the other hand, the interpretation is firmly grounded in empirical analytical method.

Extracting three Darwinian themes, variation, transmission and differential persistence, Larson et al. have constructed a three tiered meta-model of human adaptation:

> "First, individual human phenotypes vary. Second, human phenotypes are in part determined by information transmitted through two inheritance systems, genetic and cultural. Third, variation among human phenotypes leads to different levels of achievement at transmitting both genetic and cultural information through two different inheritance systems" (Larson et al. 1996, 220).

This model represents a shift of emphasis for which population growth/decline is the index. In effect population is extracted from the explanatory sequence and its data become a gauge of the relative merits of specialised or generalised economic strategies. This model may most easily be tested in extreme ecological conditions where quantifiable variables will have a powerful impact upon survival strategies.

In the evolutionary scheme, subsistence strategies may be: 1) General - usually associated with low calorific intake and hence low birth-rate/low population. However, there is an inherent flexibility which may enable survival during changes in prevailing climate. 2) Specialised - usually associated with increased calorific intake, reflected in higher birth-rate/higher population, but vulnerable to changes in conditions due to having only a restricted range of skills and experience. 3) Specialised, but "improved by tactics that promote increased subsistence stability" (Larson et al. 1996, 220-221). Any strategy is employed in the face of risk; gains in productive capacity must be balanced against vulnerability to climatic variation (Larson et al. 1996, 223).

Using variations current since the 1980s, a set of tactical models have been explored, which might be employed to implement the third strategy. These tactics might be agricultural practices, storage, exchange, sharing or warfare (Larson et al. 1996, 221).

With these models the authors were able to make reasonable predictions about human behaviour in environments where natural constraints are pronounced, and where key variables can be measured in terms of their intensity and chronology. By using climatic data broken down into months, recorded over the period 1897 to 1985, and interpreting the impact of climate on tree-rings over that time, it was possible to use the Palmer Drought Severity Index (PDSI - Palmer 1965) to make predictions about agricultural and undomesticated resource yields. The records showed that the PDSI-based model appeared to "capture more than 80 percent of the extreme events during the last 100 years" (Larson et al. 1996, 229).

Reconstruction of rainfall for the period from AD 900 to AD 1300 showed spells of consecutive dry years and spells where year to year variation occurred. Analysis of datable ceramic and chipped stone showed a "very strong association among increased climatic variability, increase in intensity of exchange, and increased dependence on storage between

AD 900 and 1150" (Larson et al. 1996, 229). Larson et al. are able to show that periods of abandonment, and evidence for nutritional stress in skeletal remains (from which "It is indisputable that reproductive rates for the Black Mesa Anasazi were compromised"), have a clear correlation with climatic changes at both Black Mesa (Larson et al. 1996, 230-231) and Vermillion Cliffs (Larson et al. 1996, 234-235). They use artefactual evidence to show adaptive strategies including increased storage, trade and aggregation, as predicted by the model.

This narrative has been achieved without setting foot in the field. It is a review of existing data, benefiting from high-resolution environmental archaeological techniques. By taking well-defined geographical locales for which tight artefactual chronologies exist, there has been a great increase in the quality of class F, which may inform understanding of class G, narrative. However, this level of precision can be achieved only very rarely with present techniques. The comparatively narrow spans of settlement activity associated with particular ceramics have obviated the need for a clear stratigraphic sequence. In northern Europe such resolution for a prehistoric period is only likely to be achieved in a wetland region with a long sequence of aggradation.

The framework for both analysis and final narrative was provided by a very distinct set of models, a feature also of the work of Bettinger and, in particular, Foley. In different ways both Foley and Larson et al. focus on the issues of land holding capacity and population; for the former a practical understanding of general population behaviour has been gained through observation of recent human behaviour in a landscape retaining resources broadly similar to those available for earlier groups. Thus a very specific population hypothesis is produced against a generalised landscape backcloth. The reverse is true of Larson et al., where tight controls over environmental features provide a sharply delineated background to population trends, rather than absolute figures.

2.3ii Thy, Denmark

Spatial archaeology depends upon an acceptable frequency of sampling and a sample size large enough to be susceptible to meaningful statistical analysis. Steinberg, from a review of the literature (Steinberg 1996, 385), has asserted that "A site that has been ploughed must have at least 20,000 flakes in a hectare to be discovered during fieldwalking". While so grand a statement is difficult to sustain the problem of the significance of the visible has to be addressed where low surface recovery rates occur. At Thy, in Denmark, Steinberg has attempted to apply a *spatial* methodology by sampling the ploughzone, sieving soil from 30cm-deep test pits.

Lacking the behavioural models offered by apt ethnoarchaeological observations from comparable ecological environments, the recovered data was grounded by excavation at selected "sites". He intended to provide "site signatures" derived from Schofield's assumption "that material remains were the end product of a given aspect of human behaviour.... whether settlement activity, quarrying or manuring" (Schofield 1991b, 5); but he hoped to offer not merely the qualitative description of surface survey, but a *"quantitative* assessment of the lithic scatters that detail the different types of lithic activities that took place" (Steinberg 1996, 375). Without a specific model the aim is to "arrive at an overall picture of behaviour that relates to stone-tool production and use.... to understand how the amount and frequency of lithic production correspond with use of stone tools" (Steinberg 1996, 375).

From analysis of the *signatures* (see *4.3iiid* below) of eighteen sites, focusing on three examples in particular, Steinberg distinguishes class F attributes such as *"ad hoc* production" (implying a degree of tool self-sufficiency), the "consumer site" (producing few artefacts and presumed to bring in many tools); and sites where production seems to have been important, for instance specialising in axes (Steinberg 1996, 385). He infers that from the Early Neolithic to the Early Bronze Age there was variation of economic function from one settlement to another, and in particular claimed that quantitative analysis has shown that in the Thy area lithic artefact "Production seems to be concentrated in a very few sites" (Steinberg 1996, 388).

2.3iii Neo-Thermal Dalmatia Project

Over the past three decades the Mediterranean has become a theatre in which the tensions between competing approaches have been played out. Chapman et al. (1996, 9-11) have attempted to assimilate a variety of approaches and meta-models to maximise the return from field data collected in Ravni Kotari, Dalmatia. The ambitious plan was to draw together landscape, social and spatial archaeologies, and "regional studies" to create a new theoretical framework, "Social Power in the Landscape" (Chapman et al. 1996, 9). Four explanatory models are introduced

"... to provide concrete explanations of the wide variety of ecological artefactual and symbolic data.... These are the land use capability model (LUC); the cyclic intensification-deintensification model (CID); the communal membership of property model (COP); and the arenas of social power model (ASP)" (Chapman et al. 1996, 252).

A strong LUC-type model requires ecological data covering the span of the period of study; reliance on knowledge of present soil distributions and climate is insufficient. By isolating different categories of land dependent resources (forage, tree crops, cereals and vegetables) and ranking their practical performances on particular soils, bearing in mind the effects of climate, Chapman et al. are able to provide an "overall ranking" of soil-type performance expected for different periods (Chapman et al. 1996, 253-258). Thus, while there is a consistent expectation that Stony land will perform well from the Neolithic up to the present (in the Neolithic it is predicted to be second preference out of 5 soil types), Terrace soils vary from being the least preferred in the Neolithic to equal first preference for the Iron Age.

In the drier Neolithic climate arable land would be first choice because it would be susceptible to cultivation; wet arable is heavy and hard to till. Therefore, the distribution on arable land of all but one of the Neolithic/Copper Age settlement sites diagnosed by the survey conforms well with the model's predictions (Chapman et al. 1996, 260).

During the Bronze Age the climate varied between warm and dry to cold and moist, increasing the risk of runoff on thinning Karst soils (Chapman et al. 1996, 255). The moistness of the arable soils would have rendered them less tractable (*Table 2.3*) and the model predicts that settlement

Table 2.3. Land Use Class capability estimtes for the Bronze Age

	Forage	Tree crops	Cereals and vegetables	Overall ranking
Arable	1	4	2	2.3
Stony	2	1	2	1.7
Terrace	4	2	4	3.3
Karst	4	2	3	3
Bottomland	4	5	4	4.7

(From Chapman et al. 1996, 255; table 32)

would favour the upper sides of valleys. The data showed a preference for settlement on the Karst, but it is argued that most of this land would have been Stony at that time. Thus, the model successfully predicts that Stony, followed by arable lands, represent favoured locations. There is no indication of pressure on land resources and the general pattern is of "a moderate degree of settlement nucleation" (Chapman et al. 1996, 261).

The data suggest that the population was much lower than the land was capable of holding, possibly due to climatic unpredictability. This issue might be better understood if more extensive palaeobotanical evidence than the very limited sample from three "test excavations" was available. However, testaments to agricultural planning are stone walled field systems.

This model is inherently unsuited to the low volume and quality of the recovered data; without a clear chronological link between particular phases of settlement and specific climatic variations it is impossible to offer a meaningful narrative. If soil hydrology is the key, there is no value in treating settlements over millennial spans as synchronic when there have been sharp fluctuations over that period. Only if close links between settlement and climatic phases can be established will we be in a position to offer explanations for certain choices of location in a landscape where resources are not under population stress. The explanatory power of Larson et al.'s models resides in their ability to provide resolution of around a decade. While the Dalmatian landscape offered a less rigorous life than the Black Mesa, we should not underestimate the frequency of need for Bronze Age groups to revise their subsistence and settlement strategies in the light of changing conditions. Unless the LUC model can be based on more highly resolved, climatic, artefactual and structural data, a narrative will be poorly founded. The assertion that "settlement distributions defined for the survey zone show broad agreement with the model's predictions" may be true, but this is a function of chronological generality on both sides of the equation (Chapman et al. 1996, 267).

The authors hoped that "discussion of the social structure" through use of the CID model would generate a more meaningful narrative. It derived from observations of "long-term social and settlement change" which appeared from John Bintliff in the early 1980s; it is the dynamic structure of the meta-model, rather than the specified socio-political assumptions, which attracted Chapman et al. (1996, 267).

The meta-model assumed that settlement alternates between expansion and contraction. Expansion will coincide with land use intensification, contraction with agricultural deintensification.

Bintliff's attempt to gauge intensity of land use "by the size, density and degree of hierarchy of settlement remains" (Chapman et al. 1996, 270) was dismissed in favour of "rather more objective measures" (*Table 2.4*). A "measure" is only as valuable as the manner of its use allows it to be: distinguishing discard patterns within and between settlements seems apt, although the "concept of settlement foci consisting of occupation sites and their 0.5 km territory" is arbitrary to the point of spuriousness. Equally spurious is the notion of "discard patterns between settlement foci" unless very tight chronological controls over artefacts can be established.

A different chronological problem undermines the use of cairns and linear features for diagnosis; while it may be possible to demonstrate a date of origin for such features (usually by associative alignment or integration with other datable structures) it is notoriously difficult to provide a date for the cessation of use. It is also unclear why tree crops, for most of prehistory only available as a wild resource, should be given priority as indicators of land use intensity; cereals represent better comparanda (albeit from limited samples) for this period, and might offer insight into trading patterns since in dried form some crops have the potential to become commodities (Jameson et al. 1994, 345). Finally the twentieth century example of massive projects being achieved by hand tools in Communist China is sufficient to indicate what volume of labour can achieve; being better placed technologically to achieve a surplus cannot be equated with gaining that surplus. Once again the column headings, referring to periods, are so general as to defuse the explanatory dynamic.

Ultimately the Neo Thermal Dalmatia Project narrative fails due to a lack of synthesis. It is, for the greater part, generated within the framework of its models, and wholly lacks the coherence and cogency of, for instance, Barker's satisfying account of the Bifemo Valley, or Foley's much more limited reconstruction of life in the Amboseli.

2.3iv Maddle Farm, Berkshire

Gaffney and Tingle's work at Maddle Farm, Berkshire, represents the most rigorous example of the application of

Table 2.4. Measures of agricultural intensification

	Neolithic - Eneolithic	Bronze Age	Iron Age	Roman
Discard within foci	low	variable	low	high
Discard between foci	low	moderate	very low	high
No. of cairns	-	137	10	27
No. of linear features	-	6	2	8
Olive/vine production	none	none	low	high
Types of plough	none	bronze ard	iron plough	iron plough

(From Chapman et al. 1996, 255; table 37)

an empiricist, problem-solving approach to regional survey carried out in Britain; it compares well with the best of recent Mediterranean surveys. Diverse but coherent techniques were supported by an elegant argument which made effective use of existing literature concerned with population and land economy in the Roman period. The rigorous standards of evidence and argument precluded the introduction of the more freely speculative reasoning which has allowed other authors to construct complex and dynamic social models.

A great deal of work was invested in lithic analysis, from which it was possible to make limited statements about the movement and preparation of flint. It also became clear that human occupation became much more widespread during the Late Neolithic and Early Bronze Age. However, there was no attempt to argue for a higher class of narrative, exploring wider issues of subsistence and social reproduction, beyond those issues closely linked to the means for obtaining flint.

Whereas previous work had failed to distinguish differential function and status of Romano-British settlement in Berkshire, the survey observed a graduation from the modest higher-status villas, through small nucleated settlement, to more dispersed (though poorly understood) instances of occupation. It is argued that the villa estate practised extensive arable agriculture from the mid first to the mid third centuries A.D., indicated by manuring scatters, as part of a cash economy which supported a "considerable surplus population, of either a civilian or a military nature" (Gaffney & Tingle 1989, 239-241). By the later Roman period there seems to have been a decrease in the extent of cultivation, a pattern which has not always been observed elsewhere. A much fuller treatment of this narrative is given in section *4.3iiib* below.

Although the narrative is clearly focused on the West Berkshire Downs, the argument is situated within the framework of what is known of Britain's economy during the Roman period. It is notable that the decline in cultivation around Maddle Farm had begun before many urban settlements reached their floruit. Within this very lucid account tenure and ownership are discussed, but there was no attempt to depict beliefs and social conditions other than at the macro-regional level (Gaffney & Tingle 1989, 242).

2.3v   South Dorset Ridgeway, Dorset

The problem-solving model as practised by Peter Woodward, with some adaptations, has influenced the pattern of regional surveys in much of southern Britain, notably the Wessex area. Shortly after completing the Great Ouse survey Woodward had the opportunity to focus on an area with an even more prolific archaeological literature when he set up the South Dorset Ridgeway Project (1977-84). Topography and ecology were given far more prominent roles (Woodward 1991, 5-11) and, with environmental evidence from excavations and cores (Woodward 1991, 127) from just beyond the survey area, they were carefully integrated with the literature and recent surface collection to form a strong synthesis. Taking into account the suitability of particular flora to particular soils and noting the effects of erosion over successive periods, Woodward presented a coherent view of a landscape evolving from the closed oak, elm and lime forest of the Late Mesolithic, when hunting is likely to have taken place, particularly in the coastal area where, at Portland, shell middens illustrate the exploitation of marine food resources.

Using the insights of Shanks and Tilley[1] to amplify and elucidate an Early Neolithic landscape of open woodland and clearings (determined by mollusc analysis), housing an increasing number of monuments from causewayed camps, to long, bank and round barrows, Woodward created an ambitious narrative. Specific ritual practices were posited (excarnation at places used for exchanging goods, preservation of relics "banked in the houses of the dead") and barrows were considered to have assumed a political role in legitimising territorial ownership. By the Late Neolithic pictograms recovered from ditch segments of the causewayed camp at Flagstones may "be taken to signify the presence of tribal groups", one of which gains precedence over two to the east and west of the area, by controlling the Frome valley (Woodward 1991,129-140). Although "lack of proven settlement sites... does not allow much scope for the interpretation of the economy" Woodward has generated a notable account of ideology at the F and G level, which informs the social narrative at similar levels, implying discrepancies in power and, consequently, hierarchy.

[1 Shanks, M. and Tilley, C. 1982. Ideology, symbolic power and ritual communication: a reinterpretation of Neolithic mortuary practices in Hodder, I (ed), *Symbolic and Structural Archaeology*, Cambridge.]

The general trend during the Neolithic was for a wooded landscape to be broken by an increasing number of clearings, some of which became broad, open swathes enabling intervisibility between monuments and testifying to the inroads of agriculture from the beginning of the period. This process continued and almost certainly intensified during the Early Bronze Age, so that by around 1400BC the landscape may be characterised as open, apart from pockets of woodland on the steeper slopes. Continuity from Neolithic traditions is exemplified by the demonstrable replacement of wooden posts with standing stones at henges. By incorporating new burial data, the significance of round barrow distributions has been reconceived. They are strung in circuits, "a major one in the central zone of the study area, and two smaller sets to the east and west [which] relate to the monuments previously defined as providing foci for the three territories established during the Neolithic period" (Woodward 1991, 141). Woodward sees a tradition, persisting over several centuries, of power in the central zone emanating from the long standing, and periodically reconceived, henge/palisade at Mount Pleasant.

Although structural remains of settlements of this period are generally lacking, geomorphological studies suggest intensification on arable use of land in the valley, causing a distinct erosion horizon, followed by a more stable period probably indicative of grazing. Eroded surfaces were sometimes sealed when pressure on land brought about the removal of woodland even from steep slopes to provide yet more cultivatable land. Evidence that this pressure led to violent rivalry might be inferred from the association of concentrations of barbed and tanged arrowheads with enclosures, in other parts of southern Britain.

By the Middle Bronze Age Mount Pleasant and other prestigious monuments had fallen into disuse and it was not unusual for them to be slighted by ploughing. A few settlements of this period within the survey area have been excavated, and point to dispersed groups of two to four enclosed and unenclosed dwellings set in or over aggregating and some coaxial, ditch-defined, plots. There is no evidence for the larger, more prestigious settlement enclosures which occur to the north of the river Frome, in central Dorset, where coaxial field systems are etched onto a previously undivided landscape. Considerable labour must have been required to sustain what would have been a mixed agriculture possibly including an element of transhumance, to limit the erosional effect of maintaining stock on the same plots of land year round.

The excavated evidence suggests that there was little rebuilding as dwellings decayed, implying mobility of population with respect to the land worked. However, it has been argued that labour would need to have been organised for the production of surpluses with which to carry on the trade evident from imported prestige goods in the later stages of the Bronze Age, so placing further stress on land resources. The creation of bounded plots would have mitigated the effects of erosion and facilitated intense manuring to replenish impoverished soils. But by the late Bronze Age "new social controls developed and began to be imposed through new types of defended enclosures" (Woodward 1991,154), culminating in the Iron Age hillforts.

2.4 The Economic Landscape

2.4i   The Dorset Cursus, Britain

Two further important surveys were carried out in Dorset during the 1980s; the field survey (and excavations on the hillfort) around Maiden Castle (Sharples 1991) and the landscape survey of Cranborne Chase (Barrett et al. 1991). The explicit theoretical perspective of the latter is the "archaeology of social reproduction". Barrett is critical of the fundamentally "static representations" of cultural and spatial models which fail to recognise the need for "information about the processes which generate those particular systems" (Barrett et al. 1991, 6) which they describe. The model espoused by the Cranborne Chase project

> "considers how people reproduce (1) their material conditions through their actions upon the environment; (2) the social system by maintaining the demands, and meeting the obligations, of social discourse; and (3) their knowledge and understanding of how to proceed in such practices" (Barrett et al. 1991, 6)

The requisite information may prove accessible because the "landscape, its form constructed from natural and artificial features, became a culturally meaningful resource through its routine occupancy" (Barrett et al. 1991, 8). The subsequent narrative reflects this concern with the physical scene; environmental data are contributions to the reconstruction of that physique. Thus the 10km long Dorset Cursus, the earlier part of which was laid out at around 2600bc, came into being in a largely regenerative woodland. It is not highly visible from a distance, but from within it the houses of the dead, already existing long barrows, exhibit their function in their alignments with the rising moon.

The presence of death is emphasised in the shape of the Cursus' southern terminals, which "imitate two nearby long barrows" (Barrett et al. 1991, 54), and is reinforced at mid-winter when the sun, viewed from the Bottlebrush terminal, sets behind the Gussage Cow Down barrow.

Although the Cursus may not quite draw a line under the temporal span of long barrow construction on the Chase, it anticipates "an important division in contemporary landuse" (Barrett et al. 1991, 54), at once illuminating the integrated and "complex design" for "those who had access to certain specific points inside the monument" (Barrett et al. 1991, 58), while separating the territory of the dead from that of the living on the lowlands. Significantly, there are more lithics of this period on the lowlands, but few monuments (Barrett et al. 1991, 54).

Despite the initial investment required for the Cursus's construction (estimated to exceed of 400,000 work hours for the totality of its length), there is no indication from the ditches that it was refurbished. Drawing on evidence from Grooved Ware associated with Down Farm, situated mid way along the north side of the Cursus, Bradley notes that "Rituals can form part of everyday life, and even if particular sites witnessed structured deposition of artefacts or animal bones, that is no basis for categorizing those sites as a whole" (Barrett et al. 1991, 83). The division of the living from the dead is less marked in the larger space, although at the settlement level certain finds share a distributional bias in the direction of the Cursus. Thus, within a domestic unit, the ritual and utilitarian co-exist, and "structured deposition" testifies to ritual behaviour which may have structured social relations. In this way the "political geography of the Earlier Neolithic" which informed those activities of self-conception at the moment when the other (in the shape of the dead) was introduced into the mundane, remained the focus of activities through which "the continuity of society was maintained" (Barrett et al. 1991, 106).

Bradley speculates on a tension between the forces for continuity between the old world of the Cursus and the long barrows, and the development of "long-distance exchange networks... a factor in the rise of Hengistbury Head", which "may account for the change in political geography marked by growth of the Knowlton complex". Unfortunately, little fieldwork has been carried out on this group of hengiform circles (Barrett et al. 1991, 83), about 4km south of the Cursus, beside the River Allen, but the location is symptomatic of a trend towards significant occupation along the larger rivers (Barrett et al. 1991, 107). By the end of the Neolithic the shift of domestic activity away from The Chase is demonstrated by increased vegetation cover.

A graduated sequence of change in ritual behaviour through the Later Neolithic is illustrated by the Wor Barrow complex, about 1km northwest of the Cursus. In the earlier long barrow itself, the presence of unarticulated and later articulated human skeletons enshrines a change in practice, and further transformation of tradition towards focus on the individual is exemplified by the initially single burials which were sealed below the mounds of the two round barrows (also Neolithic, but later) oriented with respect to the long barrow.

Barrett stresses what he considers to be the brevity of the process of inhumation and notes consequences: if grave goods were included, then their prime significance did not reside in their prestige value since their visible association with the dead person would have been curtailed. Thus goods were simply symbols of the "statuses and obligations which were claimed from the individual by the living"; in other words, grave goods represent the routine daily activities in which the person had been engaged during life. As such they are an affirmation of that individual as conceived by his/her community. Barrett sees these simple inhumations as expressions of a "shortlived and localised political aim" (Barrett et al. 1991, 223); however, in a society where eternal verities no longer resided in the ancestors, the naked fear of ceasing to be might have been assuaged by the identification of individual with place - particularly the medium of source and nurture, the soil.

Barrett exhorts us not to view this as "hierarchy or centralisation within the social system, but as arising from the transformation of social practices" (Barrett et al. 1991, 138). The two are surely not mutually exclusive. Whilst one may readily accept that the new rites of passage for the dead represent a shift from a belief system dominated by the ancestors in general, to a more particularised "classification of the dead" as individuals (Barrett et al. 1991, 139), one is forced to consider what is implied by the "restructuring of inheritance amongst the living"; exclusivity and hierarchy seem at least implicit. If, indeed, the structural roots of the Middle Bronze Age's organised agriculture are to be found in the Early Bronze Age's systems of kinship and labour obligations, exemplified in the differentiation within and between death rites, the wealth of barrow forms and the focus of certain categories in particular areas surely suggests that differing levels of authority were being passed on. Unfortunately, Cranborne Chase, as for most of Britain, offers restricted evidence for the activities of the living other than those precipitated by the "temporal irregularity" of death.

Cranborne Chase's Middle Bronze Age emerges as the 'Wessex' barrow tradition closes, a period when the routine activities of the living are encapsulated in field systems and enclosed, as well as unenclosed, settlements. The Cursus remains a significant seasonal boundary between up and lowland patterns of activity in a "landscape structured around the seasons of agricultural reproduction and the cycles of animal migration" (Barrett et al. 1991, 223). The characteristic artefact of this period is the pot, in particular, the increasingly mundane Deverel-Rimbury form. Excavations at Down Farm, Martin Down and, possibly, South Lodge revealed enclosures constructed after earlier buildings, and they are sometimes demonstrably later than field systems. The ditches enclosing South Lodge cut lynchets, so illustrating the sustained regularity of cultivation of particular plots of land and, with that, the growing capacity of humans to control their environment.

The burials of the Early Bronze Age, where the monuments of the dead become more elaborate to accommodate "increasing concern with particular lines of inheritance" and the immediate political crisis of authority in transition after a death, contrast with the scale of Middle Bronze Age monuments. Where there is direct continuity of use, the later burials are inserted into the peripheries or ditches of barrows; if new barrows are created, they are smaller. Death itself has become mundane, with cemeteries situated close to the settlements of the living, where previously they had been at a distance. The dead go to their graves with storage and cooking vessels, so that the raw products, and their preparation as food, symbolise a continuity; in this, seasonal cycles and reproduction have become the dominant ideology (Barrett et al. 1991, 225).

It may be apt to characterise the Wessex Middle Bronze Age as a period of stability within which humanity greatly extended it influence over the landscape. The events of the succeeding Late Bronze Age, a phase which includes some defensible enclosure construction, have been explained as either a consequence of population growth associated with the success of agriculture or as a result of extrinsic factors involving trade or gift exchange of exotic items. In either or both cases discontinuity is a recurrent attribute of the settlement pattern. Abandonment of settlement in the Bronze Age should be seen as "the renegotiation of age and gender status and of political alliances based on kinship" (Barrett et al. 1991, 239). These are moments of closure, rather than rupture, within a recognised discourse. But in the Late Bronze Age rupture seems an appropriate theme.

Reorganisation of space occurs within settlement (for instance, a large rectangular building at Down Farm, implying "a radical reorganisation of the domestic space") and within the landscape, where a perceived lack of social and economic autonomy led to a "break in the settlement sequence as we currently understand it, with settlements reappearing on a number of new locations by the seventh century BC" (Barrett et al. 1991, 239).

The pattern of small hoard deposition (contrasting with the substantial deposits in Fenland Britain) gives way to finds of spearheads and axes "in isolation or in hoards" in the Late Bronze Age, in the Wessex region. Rowlands (1976) has offered models to explain the pattern based on barter, equal gift exchange and competitive giving (Barrett et al. 1991, 239) but the small scale of Wessex metalwork distribution suggests that the explanation for their deposition is likely to rest in the mundane, rather than in competition. The record suggests that competitive display seems more likely to have featured in feasting (Barrett et al. 1991, 240); independent evidence of this has emerged at the Wiltshire midden sites of Potterne (Lawson 2000) and, since the Cranborne publication, East Chisenbury (Brown et al. 1994; McOmish 1996).

The enclosed settlements of the seventh century BC, and the buildings within them, were on a larger scale. There was a "greater range of agricultural storage facilities and they seem to have had a longer duration, lasting well into, sometimes throughout, the Iron Age. Settlements were likely to have been integrated into the linear boundary ditch systems which in southern and eastern England define "large tracts of land, often on the boundary between major changes in soil types, or rivers and coastal havens... characteristic of major administrative and political centres or 'oppida'" (Barrett et al. 1991, 240).

It is argued that in the study area a "sub-Durotrigian political entity" held sway "through import of exotica and finally through the direct patronage of Roman military and provincial administration itself (Barrett et al. 1991, 242). They may have adopted from Gaul a system of dual magisterial authority, which facilitated a comparatively smooth transition to a Roman rule which, devolved via native authority, took its place within larger systems of political reproduction" (Barrett et al. 1991, 242).

2.5 Socio-economy: states of mind

In this section I shall outline the narratives of two major Mediterranean surveys. Two further surveys from that macro-region are dealt with in detail in *4.3iiic* and *4.4iic*. As with the previous section I have employed an "earliest" cut-off date, going no further back than the Neolithic, and a latest, which may be described as Early Medieval. These choices, particularly perhaps the latter, may seem arbitrary and to run against the increasing trend to include all periods represented by material residues. Aston (Aston & Costen 1994, 19) has expressed the view that by working backwards one can effectively undress the landscape. In addition, his colleague Chris Gerrard's legitimate concern is that Late or Post Medieval artefacts may be discarded as uninteresting; his fears are well-founded (Aston & Gerrard 1995, 13). In the otherwise rigorous Tarragona regional survey "any cultural material" was collected, but everything considered to be Post Medieval was thrown away once a collection unit was completed, or during finds processing (Carreté et al. 1995, 52). There may be some force to the argument that further cultivation will turn up more, similar material, but it has been demonstrated that in some locations high volume erosion removes artefacts as well as soil (Bell & Boardman 1992, 233-234). Their relocation remains of geomorphological interest, but surface distributions may no longer occur close to the place of original discard, and with that a particular resource will be lost.

My objective is to consider how certain classes of narrative may be attained from comparatively restricted data. In this respect the admission of historical material is inappropriate, except where it provides a crude rule against which to measure archaeologically based accounts. Yet as will be apparent from the narratives in this section, data from later, text informed periods and, crucially, interpretations, are used to shape the perspectives and models within which prehistoric activities are framed.

2.5i   *Tarragona hinterland, Catalunya*

The most recently completed survey of the four was carried out from 1985 to 1990 in the hinterland of Tarragona, Catalunya. In essence it follows a problem-solving paradigm, introducing models on the basis of survey-generated data. The questions posed are of a sophistication which demanded a quite different, multi-layered, approach to that of Woodward's earlier work (1978), and specific acknowledgement of Foley's off-site perspective (Carreté 1995, 46) demonstrates a keen theoretical awareness. The narrative presented in the current volume is designed to stand alone in dealing with a particular time span, from the Bronze Age/Iberian transition, to the end of the Late Imperial period. A second volume is due from Schofield et al., dealing with earlier prehistory.

The survey area comprises over 600 square kilometres of land ranging from the 300m asl foothills of the Serralada Pre-litoral in the north, to the coast in the south. To the east it is bound by the Jurassic limestones of the Bonansa Massif, to the west of which the river Gaià flows southwards. Parallel with it, to the west of centre, the study area is bisected by the river Francolí, the alluvial silts of which have long provided much of the better arable land in the region.

The climate is classically Mediterranean, with a maximum summer mean temperature of 24° in July, and a winter minimum average of 9° in January. Mean annual rainfall is within the range of 600-800mm, less than 20mm falls in each of June and July. Autumn is the wettest time of year, with a peak average of 90mm in September. Violent late summer storms can cause local flooding, rapid erosion, and down-cutting in the valleys (Carreté 1995, 39-41).

The alluvial deposits along the main river valleys vary in depth, but are generally thicker towards the coast. A recent study (Corbertera 1986) of the present soils in much of the region shows that brown earths in the central, west and north parts of the area provide moderate and some good arable land, but that further east the solution weathering of limestone brings upward migration of calcium carbonate in the high summer temperatures, forming a crust which limits the lands' agricultural potential. Such land is usually under *maqui*-type scrub.

The problems selected for the archaeological programme were period-specific:

1) What was the degree of rupture of the settlement pattern during the transition from the Bronze Age to the Iberian Iron Age;

2) What was

   a) the density and character of Iberian settlement;

b) the relationship of the region's rural periphery to the developing urban centre;

c) the index of agricultural intensification;

3) Do changes in the rural areas correspond with the emergence of Tarraco as a strategic centre during the Republican period in terms of

a) the ratio of abandoned Iberian and new Republican period settlements;

b) the density and preferred locations of settlement;

c) contact with Tarraco represented by imports recovered from the hinterland;

4) Is there a trickle-down of Tarraco's wealth and power to the hinterland?

a) What evidence is there for rural nucleation and ownership by a Tarraco-based élite;

b) How focal did Tarraco become;

c) Is there an increased density of use of marginal land;

d) To what degree did the territory become self-sufficient in cash crops;

5) Is the urban decline of the Late Imperial period reflected in a rural renaissance?

a) Did Tarraco continue to influence the location of high status sites;

b) Did nucleation persist;

c) Did rural communities survive the invasions of the third and fifth centuries AD.;

d) To what extent did Tarraco serve as a centre for the redistribution of imports (Carreté 1995, 45-46)?

The first problem is confined to class E and F-type questions; but class E analyses are clearly destined to address class F and G problems. The framed questions concern the *social* narrative classes, but in the event the narrative does include some class G *ideology*.

A second, and fundamental, concern of the report is the disparity in levels and quality of information that has been presented from survey to survey. In part this is a problem with the evolution of strategies, but the persistence of different perspectives has created an enduring problem: the lack of a common language for comparison, and great difficulties for those seeking to synthesise survey information, as noted by Alcock (Alcock 1993, 35-36; see also Alcock & Cherry 1996, 210) in her admirable attempt to harmonise the modern documentation of Roman Greece.

The survey failed to recover Bronze Age data, making it impossible to pass informed comment on the transition from that period to the Iberian Iron Age, but it is noted that Schofield et al. (forthcoming) will show that "the area was intensively exploited by the Neolithic period", and that it is unlikely to have been deserted thereafter, until the widespread settlement growth of the 5th century BC. Given that there must surely have been Bronze Age settlement the methods of the survey require scrutiny.

From the Iberian period onwards the report is equipped to present class E, F and G narratives. Using a much subtler approach to ceramic distributional analysis than has occurred elsewhere the authors have been able to show that by the Iberian period it is possible to distinguish supply zones for local fabrics, against the background of an efficient trading network, illustrated by the widespread occurrence of Punic imports.

To the two major sites of this period may be added a third, site 3.19 (Carreté 1995, 201). Earlier excavations uncovered contemporary Greek pottery from modern Tarraco and El Vilar which has not been found elsewhere in the area; the authors speculate that the substantial settlement situated on a dominant hilltop to the east of the Francolí, some 15km north of Tarragona, might reveal similar material if it, too, was excavated (Carreté 1995, 275-276).

Carreté et al. propose that El Vilar and 3.19, both perhaps pottery production centres commanding routes along the Francolí valley, and with access to varied agricultural resources, may have played key roles, comparable with, or slightly less influential than, the settlement situated at modern Tarragona. The latter may have gained ascendancy because in addition to good agricultural resources its coastal location gave immediate access to exchange networks, as well as to sea food. The full hierarchy would have included specialist production settlements and farmsteads of varying size dispersed at roughly one to every square kilometre (Carreté 1995, 275-276). The only immediately obvious difficulty with this account arises from the relative proximity of El Vilar to 3.19, at less than 5km, whereas those two sites are at approximately 16 and 15km respectively from the site at Tarragona. At the very least this suggests a pronounced discrepancy between the influence of the Tarragona site and the other two; alternatively, there may be major sites which have yet to be found.

One of the most important aspects of the report is its effort to represent, if not transition itself, then at least the degree of continuity of site-specific settlement (Carreté 1995, 247-250; Figs. 9.5 and 9.6). Thus although it is accepted on the basis of textual evidence that the creation of a Roman supply base at Tarraco, heralding the region's Republican phase, "stimulated trade in the region" (Carreté 1995, 276) which, the evidence suggests, "distorted and then dominated the surrounding area" (Carreté 1995, 265), the frequency of settlement remained broadly similar. Indeed, the great majority of sites remained in use.

The military presence would have attracted Italian traders, whose wares testify to the swift and widespread growth of an effective trade network which reached most settlements in the region, regardless of their status and distance from the port. Although inland centres continued to be important for the redistribution of local products, Tarraco's pre-eminence as the principal axis of supply by the late 1ˢᵗ century BC is marked by the abandonment of El Vilar, and the lack of comparable growth at 3.19.

Settlement density continued at around one per square kilometre, suggesting that by the end of the Iberian period the land was close to its holding capacity, presuming that average settlement population group-size remained unchanged. Where new Roman farmsteads occurred they did so alongside native farms; there is no evidence that the land was being centuriated or made available systematically to veterans or other colonisers (Carreté 1995, 276).

During the succeeding Early Imperial phase Tarraco grew to cover approximately 70ha, close in size to Gaul's larger provincial capitals. About 25% of this area was given over to public buildings, an unusually large proportion. The authors estimate that the population lay between 10,000 and 15,000, constituting upto 20% of the people living in the territory, the remaining 66,000 of whom were distributed in an estimated 3,300 rural settlements (Carreté 1995, 277-278).

Repeating a pattern found elsewhere in the Mediterranean at this period, Tarraco appears to have had a green-belt radiating approximately 4km around it, perhaps providing a horticultural resource for the town. But, as the preponderance of public buildings implies, it was a political, rather than a market, centre, to which the ambitious gravitated, taking up residence in the high status villas on the coastal strip, or on estates in the lower Francolí valley.

The densest rural settlement was beyond the green-belt, but still within easy reach of the city. More high status villas occupied this area, although its agricultural productive potential may have been as great an attraction as proximity to Tarraco. The continued effectiveness of the trade network is exemplified by the greater range of finewares (Carreté 1995, 260-261) found in all parts of the region, reaching all categories of site, although with a pronounced trend in favour of higher status settlements. Significantly, large storage vessels were usually locally produced, suggesting the territory's increased self-sufficiency in food.

The characteristic settlement pattern changed little, although one cluster of sites (centred on 1.6; Carreté 1995, 170) and a small area which geophysical survey showed to have accommodated several buildings (Carreté 1995, 216-221) may represent some degree of nucleation in the face of a generally dispersed pattern of rural settlement.

During the Late Imperial phase there seems to have been a gradual increase in the trend towards nucleation. There was a pronounced fall in the overall number of settlements, although the authors are at pains to point out that the low density of finds from this period have a powerful distorting effect, and may be due to the less widespread use of ceramics, as much as to a decline in settlement. Tarraco, itself, may have lost some of its population to major rural settlements with in a band radiating some 10km around it. This band probably represented the limit of supply of imported wares in the north direction, and is indicative of contracted trade networks.

Imports continued to reach this privileged area until the late sixth or early seventh century AD, suggesting that it was little affected by the upheavals associated with the Visigothic invasions of the third and fifth centuries (Carreté 1995, 280-281). During the ensuing period all Late Imperial settlements seem to have been abandoned and, eventually, replaced by the Late Medieval settlement pattern, under which some are likely to be concealed.

The report makes extensive use of excavated evidence largely derived from one locality (the city of Tarraco) to bring to life particular nuances of political life in Imperial Tarraco and, consequently, is able to show how systems of belief were manipulated to serve the interests of individuals and groups. Equally effective is the analysis and interpretation of negative evidence. In some cases a lack of surface finds coincides with a particular geology, suggesting that physical reasons determined the nature of agricultural activity there. In contrast a cluster of "blank" fields on some of the best arable land in the territory, a little over 10km north of Tarraco, seem to have been part of a previously known villa's estate; they must have been too far from dwellings to have incorporated domestic waste with any manure that was spread over the surface. Five kilometres further north a cluster of settlements is noted to lie between good quality arable land, from which surface artefacts were collected, and land less suited to cultivation, with no surface finds (Carreté 1995, 272). This suggests a planned allocation of varied land resources, shared between a number of farmsteads.

Other areas of low or zero artefact density are variously interpreted as places of little settlement, as non-arable agriculture or as simply non-productive. A striking finding of the survey throughout the periods of densest occupation was the relatively indiscriminate choice of all but the worst soil types. Topography was a factor; there was a marked preference for elevated locations, often on the slightest of rises in otherwise flat landscapes, or at the breaks in slopes.

Detailed environmental reconstruction of the kind used by Foley, Bettinger or Woodward, might have shed more light on settlement distribution, although the use of a recent survey of the modern soils and land use (Carreté 1995, 39-44) suggests that the authors were prepared to accept that the landscape had changed little since the study period. This may be the case, but a properly interpreted geomorphological survey would have enabled a greater degree of confidence in the reasons for preference and might have enhanced other aspects of what is, in all other respects, a well constructed and well founded narrative.

2.5ii   North Keos, Cyclades

Cherry et al. had aspired to a survey of the whole of the cycladic island of Keos. However, restrictions on the amount of time allowed for fieldwork frustrated that intention, thwarting the attempt to study an area which, as an island, has dearly defined physical boundaries.

The northern end of the island is 12km from west of the Attic mainland. The landscape is characterised by steep mountains, rising to 561m, and an indented coastline providing good anchorages. The study covered an 18km^2 area of the northwest of the island, cut into two distinct geographical units by the ridge of "Roukouas and the lower hills of the Fotimari" area. In general the island is well watered, and the lack of flat land is compensated for by rich alluvial deposits. In some areas ridge tops are almost devoid of soil, but slopes are or have been cultivated through the construction of terrace which are characteristic of most of the island (Cherry et al. 1991, 57-59).

To facilitate comparability the survey was conducted within a spatial paradigm, although in a slightly uncomfortable manner since the target population comprised sites, while the notion of *off*-site was treated as a distinct issue. This hybrid scheme was to accommodate two distinct classes of positive evidence (portable and structural) and negative evidence.

The interpretation of early periods of activity in the area is principally confined to observations of distribution. A preference for coastal regions is noted from the Late Neolithic through into the Early Bronze Age. The most detailed evidence from these periods came from excavations and subsequent detailed collections at the Final Neolithic site of Kephala (Cherry et al. 1991, 267-268), a small, possibly nucleated, settlement and a site of slightly later origin at Ayia Irini.

Intervisibility with the mainland, good soils and readily available water not only attracted settlers, but seems to have encouraged continuity of settlement at Ayia Irini at least. However, finds more than a short distance away from Ayia Irini are disparate, suggesting that a strong tendency towards nucleation persisted throughout the Keian Bronze Age (Cherry et al. 1991, 228). The correspondence between the presumed rise and fall of the settlement's population from its Final Neolithic foundation through to the Late Cycladic and the density of pottery in the surrounding area is significant enough to be reported, but remains unexplained (Cherry et al. 1991, 229). It is suggested that growth of the nucleated population during the Middle Cycladic, upto the second phase of the Late Cycladic may have "increased the demand for surplus" which might have required a "larger number of agricultural facilities outside the primate centre" (Cherry et al. 1991, 227). This growth may have reflected the area's, and in particular, the "primate centre's" developing integration into regional exchange systems, a pattern repeated from a slightly later date at Phylakopi, Melos (Cherry et al. 1991, 231), although the findings from survey of that island have been interpreted as showing a much denser nucleated population (Cherry et al. 1991, 221-223).

There is a distinct rupture in the latter stages of the Late Cycladic represented by a major destructive event at Ayia Irini, but the evidence is unable to detect whether this is merely signifying the culmination of a gradual loss of importance of the settlement, or the deliberate depopulation of it, perhaps due to outside agency.

The latter Bronze Age "Dark Age" seems to have cast a long shadow over the Keian landscape and its only in the Archaic period that real signs of regeneration emerge. Nucleation, this time centred on the nascent *polis* of Koressos, remains characteristic of settlement distribution, but increased rural discard rates suggest there may have been a few dispersed farmsteads, perhaps belonging to the farmers who alternated between living in the central settlement and rural periphery (Cherry et al. 1991, 230, 346).

From the 6th to the 4th century BC the agriculturally productive population would have had to respond to increased "demands of the state for surplus" (Cherry et al. 1991, 339), to make provision for "major civic construction projects". It is not clear how this increased productivity was achieved, but options included an enlarged labour force; more frequent use of agricultural land; more family land-units; a general increase in the amount of land brought under cultivation or more permanent settlement of the land. However, Alcock reminds us that "Civic grandeur can easily reflect increasing exploitation of the population, rather than a rise in *per capita* income, general prosperity or higher urban population levels" (Alcock 1993, 102). It is likely that production was unspecialized, with each unit producing a range of crops.

The system would have been based on individual or family ownership. The decline in the number of surface finds of the subsequent Hellenistic period need not represent a corresponding decline in agricultural activity. Scatters resulting from manuring can only occur when domestic refuse is incorporated with animal/human excreta, strongly suggesting that stock were enclosed in proximity to areas of human habitation. Therefore there are likely to be few residues where land has been farmed as part of an estate at a distance from the secernent The evidence from Keos is interpreted as marking a trend towards the replacement of private land holding in favour of efficiency savings such as monoculture and reduced investment in the construction of housing or other buildings (Cherry et al. 1991, 346).

2.5iii   The Southern Argolid, Peloponnese

The great strength of the report of work in the Southern Argolid, in the Peloponnese, is the manner in which environment, archaeological and historical perspectives are used to amplify each other.

The project area, some 225m^2 of a peninsula, comprises a landscape more than a third of which is over 600m above sea level, with high plateaus of hard igneous rock, some of which descend directly into the sea. Steep valleys cutting deeply through sandstone, and lower land of rolling hills, rarely exceeds 100m above sea level. Modern vegetation includes dense pine on northern slopes and a *maqui* of wild pistachio, olive, holm, grasses and legumes.

The report portrays the evolution of the landscape from its structural formation in geological terms, through the effects of sharp variations in sea level, and variations in climate. By the Early Neolithic the decrease in tree pollen and comparative increase in wood-edge herbs suggests the initial clearance of the deciduous oak open forest that had grown up in the post-glacial period. Use of plants is visible in the record from about 10,000BP, and over the next two millennia cultivated grains and lentils appear. Later in the Neolithic a landscape including evergreen oak may have resulted from the browsing of domesticates, and by the Early Helladic the aspect is one of open woodland (Jameson et al. 1994, 166-169), symptomatic of extensive upland clearance. The level of cultivation was such that by the latter part of the Early Bronze Age there was a "major phase of soil erosion" (Jameson et al. 1994, 192).

It would appear that the early farmers were aware of the problem of erosion; better soil management during the Middle and Late Bronze Age seems to have arrested its downward progress. By the beginning of the first millennium BC there are very few signs of any activity in the landscape which, to all intents, appears to have been abandoned.

From the Classical, through to the Late Roman period there are intermittent phases of erosion and alluviation which appear to coincide with local economic trends (Jameson et al. 1994, 192). Thereafter, there was increasing aridity and the evidence points to a decline in agricultural activity.

Prior to the arrival of agriculture, ethnographic evidence set against environmental reconstruction suggests that no more than 30 people would have been supported by hunter-gatherer subsistence. The occurrence of Near Eastern cultigens contemporary with the village-like settlement at Franchthi cave gives support to the view that agriculture was an intrusive phenomenon. The peninsular did not possess high-yield land, but it had the advantage that, being thinly populated, there was little competition for space.

As early as the 7th millennium BC querns were imported for the processing of wheat and barley. Meat and milk were supplied by keeping flocks of sheep and goats, and shellfish and fish were eaten, also (Jameson et al. 1994, 341). Apart from querns, flint and obsidian were also imported by sea, in exchange for foods which could be preserved by dry storage. These imported cultigens had replaced indigenous, potentially cultivatable plants, precisely for their commodity value (Jameson et al. 1994, 345).

For much of the Neolithic settlement was dispersed and very thin, with the exception of a village which, perhaps with a population of 100 shortly after its foundation, may have grown to house as many as 500 people by the end of the Neolithic (Jameson et al. 1994, 347-348). A pattern of cultivation in spring-fed meadowland, and probably some transhumance, may have been altered as rises in the sea level in the Final Neolithic were enough to cause the loss of significant areas of the coastal plain. Movement upslope would have forced reliance upon the vicissitudes of unpredictable rainfall (Jameson et al. 1994, 347). At this time there are at least 7 new settlements and the peninsular's population may have risen from around 300 to 1000.

The pattern of mid-valley settlement persisted, along with the settlement of rolling, coastal hills, into the Bronze Age. From the 5th to the 3rd millennium BC there was a massive increase in alluvial deposition, often taking the form of a phase of sudden aggradation, although the general reduction of high ground soil depth appears not to have been great (Jameson et al. 1994, 353). In an effort to mitigate the effects of sometimes low rainfall open fields were created, which would have required a larger labour force.

During the Early Helladic Bronze Age there was a pronounced move towards specialisation of production, coinciding with an increase in the number of imported items, including obsidian, flint, andesite, marble, precious metals, textiles, pottery and weapons (Jameson et al. 1994, 356). At least some of the obsidian arrived as raw material to be fashioned into prestigious artefacts at the specialist production centre of Fournoi (Jameson et al. 1994, 363).

The arrival of metallurgy in the fourth millennium, and improvements in marine transport at the beginning of the third millennium, will have engendered differential access to communication and technology. Those most equipped to benefit, whether by accidents of geography, or by previously acknowledged status within a community, began to exercise more far reaching control over individuals and communities. At Fournoi, the specialist craftsmen, who produced obsidian artefacts recognised as prestigious by comparison with other lithic objects, may have been labour commodities to be exchanged; if not, then their products at least were controlled by "elite decision makers" (Jameson et al. 1994, 363). Through this elite the Keian commodities of "salt, dried fish, honey, wax, wood, charcoal, resin, talc, hides, wool, hair, cheese, grain", and even human labour, would have entered the exchange network (Jameson et al. 1994, 356).

The picture presented for the Helladic II phase (Early Bronze Age) is of a group of "politically and socially centralized" communities or chiefdoms of nucleated and dispersed settlements (Jameson et al. 1994, 363). Some major settlements possessed querns and appear to have had more pottery in circulation, while further afield the landscape was interspersed with structures linked with farmsteads, storage facilities, animal folds, field houses, beekeeping and more besides.

The patterns of Early Helladic III and Middle Helladic activity (2500 to 1600 cal BC) are presumed largely lost to erosion. The only good quality data have come from excavations at "village" sites of Lerna and Asine, whose livelihoods were based on the continuation of animal husbandry, cereal production, hunting and gathering and, for the first time, domesticated vine and olive (Jameson et al. 1994, 367). Life appears to have been peaceful, even modestly prosperous, although by the latter part of the period there was "growing social complexity". By the Late Helladic (1600 to 1000 cal BC) the evidence suggests that there was "a widely dispersed, hierarchical settlement pattern" similar to, and often situated at, Early Bronze Age settlement areas. The presence of high quality Mycenaean sherds is indicative that certain sites became focal. The general lack of finds attributable to Late Helladic IIIc seems to tie in with a period of regional decline following the collapse of the Mycenaean palaces.

This dearth of surface pottery continues into the Proto and Early Geometric phases, suggesting large scale abandonment of "most of the land that had been cultivated in the Mycenaean period" (Jameson et al. 1994, 373), although the evidence of relative prosperity from neighbouring Attica and Argeia make it unlikely that the peninsula was wholly deserted. However, by the Middle Geometric phase settlements are reappearing on good cereal-producing lands, generally set inland. This preference may have been due to the submerging of the springline by the sea, but the few settlements near the coast are in higher, defensible positions, suggesting anticipation of seaborne attack. The wealthier farmers are presumed to have taken up residence in the larger villages, along with "the few specialist craftsmen and tradesmen" the smaller surrounding villages and hamlets could support (Jameson et al. 1994, 375). The early isolated structures were either shepherds' cotts or shrines. The latter, taken with the size and pattern of larger settlements, suggest that city-states were beginning to emerge from the earlier pattern of chiefdoms. By the 7th century BC five such poleis had developed (Jameson et al. 1994, 374-375).

The elements of this new hierarchy depended upon the ratio of productive land to the total territorial area, and upon their locations with respect to trade and administration (Jameson et al. 1994, 376). In the Southern Argolid an inward perspective is reflected in the emergence of Hermion, in the middle of the peninsula, as the dominant polls. By the archaic period the other polities had become its subordinates (Jameson et al. 1994, 377-378). Documentary evidence from the 5th century confirms that the Southern Argolid had indeed become a backwater, but its coastline would have serviced passing trade, and a low-level of importing and exporting would have persisted (Jameson et al. 1994, 380).

The classical period was one of increasing instability for a piece of land caught between the conflicting interests of Sparta, Athens and Argos. An alliance with the former led to sackings by each of the latter. Nonetheless, the majority of sites remained occupied, and there was an increase in the number of small rural sites (Jameson et al. 1994, 382).

This last detail in the pattern of settlement became the characteristic trend of the Late Classical/Early Hellenistic transition of the later 4th and earlier 3rd centuries BC. Over half of the nucleated classical sites ceased to be occupied, many after centuries of use, and others were reduced in size. One category of site, the cult centre or shrine, became obsolete. At the same time there is a proliferation of small, dispersed rural sites (Jameson et al. 1994, 383).

Although this pattern represents an increase in rural population, it conceals a larger total of people linked to the land. The discovery of olive presses in the town as well as the country, coupled with epigraphic evidence, shows that owners remained in the major centres, leasing out the farm dwellings, which may well have been in year-round use by tenants (Jameson et al. 1994, 386). It is likely that one family living in a nucleated settlement would have owned several farmsteads, which by this period were larger than in the classical period, and were tending to specialise particularly in olives, figs and vines (Jameson et al. 1994, 391 & 394).

In Greek cities generally, during the later Hellenistic and up to the Middle Roman period, the divide between rich and poor increased. The latter had to compete with the influx of slaves and immigrants, placing them in an invidious situation in the labour/production market. The rich were few in number, for this is a period of general decline, perhaps in part precipitated by piracy and brigandage of the latter 3rd and first half of the 2nd century BC. Those dispersed farmsteads which remained in use near the coast tended to have towered buildings, at once enhancing visibility and defensibility.

Halieis, Hermion's main satellite, is known to have been abandoned as early as 280BC, perhaps as the water supply deteriorated, but also probably because it lacked a powerful patron, trading conditions became unfavourable, and with a reducing population it lacked the manpower to farm intensively. Nearly all the rural settlements in its vicinity also succumbed. Hermion, itself, survived, and with it some 80 of the surrounding farmsteads (Jameson et al. 1994, 394-397).

The general decline of rural population and labour is reflected in a phase of increased erosion. Modern parallels suggest that as land ceased to be managed for crops and was abandoned or left for grazing, there was a loss of the shrubs which bound the soil, whilst animal tracks began to breakdown terraces and to create channels which were incised still further after heavy rainfalls.

By the 5th century AD there had been a strong economic recovery. Nearly half of the Late Classical/Early Hellenistic sites were re-occupied. Although Hermion remained the only town more general prosperity was confirmed by an increase from 6 to 16 nucleated settlements, and from 17 to 82 relatively isolated rural sites within the surveyed areas. The recovery may have been linked to a revived olive industry, but new industries included tile and pottery manufacture, the latter of large storage vessels, possibly to hold native produce (Jameson et al. 1994, 401-403). The scale of landholding varied from the small private farmstead to the villa estate, the greater part of whose labour force would have been slaves and land-tied freed men, as well as some tenants (Jameson et al. 1994, 401-403).

This comparatively affluent spell came to an end with general instability which followed the collapse of the Roman Empire. With trade networks disrupted the population number fell, and those remaining returned to an inland, pastoralist, way of life (Jameson et al. 1994, 404).

2.5iv   The Biferno Valley, Molise

Graeme Barker, assuming a Braudelian perspective, situates the people in the landscape, emphasising the point by including photographs of peasants at work in the fields in the 1950s. The study area, in the Molise of central southern Italy, selected precisely because it had received little previous attention, was defined by a box of 75km by 30km. It included five basins bounded by the 2000m asl Matese mountains, the watershed of the River Biferno and its main tributary, the Cigno, following the Biferno Valley to the Adriatic, including most of its 1311 square kilometres catchment area, and that of a small tributary feeding the river from the north side of the lower valley (Barker 1995, 17).

The karstic floors of the five intermontane basins are at around 1000m asl. Streams run into the Boiano Basin, at the north edge of which they form the River Biferno proper (Barker 1995, 20). The river, which has carved a valley through the mountains, flows through some 50km of soft sands and clays in the middle valley, falling from 450m asl to around 90m asl at Ponte Liscione, where it has been dammed to form the 8km long Lake Guardialfiera. Below the dam it "meanders across a wide floodplain, reaching the sea near Termoli" (Barker 1995, 22-23), etching its path through marine sands.

The modern climate varies sharply from the upper, southern, end of the valley to the lower, northern end. During the coldest part of Winter temperatures at Campobasso, in the upper middle valley, average 3°c, but they reach 7° at Termoli. In Summer the average temperatures are 22° and 25° respectively, with coastal temperatures as high in the Spring as those of the upper valley are in the early Summer. Annual rainfall averages 1870mm in the upper valley, but only 644mm in the lower. Throughout the valley the highest rainfall is in the Autumn and Winter (Barker 1995, 28).

After the last glacial period higher temperatures had been accompanied by low rainfall at low altitude and the advance of vegetation was fairly slow, consisting of open mixed oak forest and grasses on freely draining soils. Increase in rainfall from around 7500BP brought about a much more extensively and densely forested landscape, shortly before the arrival of cultivation in central Italy, including Apulia, immediately to the south of Molise. The new mode of subsistence adopted by its neighbour failed to impinge on the Biferno Valley for another millennium, perhaps because the crops favoured by the Apulians were ill-suited to the valley's soils.

Favourable consideration is given to a particular model of the transition from hunting and gathering to farming for northern Europe (Zvelebil & Rowley-Conwy 1984). These are the:

1)   "availability phase... foragers are in contact with farmers, and may obtain cultigens from time to time through trade (or by force) but make little

use of them;

2) "substitution phase... foragers make increasing use of animal and/or plant husbandry at the expense of their traditional way of life;

3) "consolidation phase... husbandry becomes the predominant mode of subsistence" (Barker 1995, 98).

In the Early Neolithic extended families occupied fairly isolated farmsteads, relied heavily on the meat of nearly mature sheep/goats (the two species are not easily distinguished), and some pig and cattle to supplement their cereal diet (Barker 1995, 113; 127-128). Although organisation was limited, prestige giving or product exchange would have helped to establish networks of mutual support necessary to sustain the population through less productive periods. For the thinly spread population, although economic activity may have structured aspects of the decision-making process, ritual activities served "as a mechanism for maintaining social boundaries and legitimating traditional authority" (Barker 1995, 127-128). However, for the sake of "health, fertility and economic welfare" there would have been a need for fluid group membership.

By the Middle, and continuing into the Late Neolithic, mixed farming predominates throughout the Italian peninsular, generally from larger, nucleated settlements. Domesticated animal were allowed to live longer, and finds such as strainer sherds and spindlewhorls confirm that secondary products such as milk and wool were more important than in the earlier phase (Barker 1995, 113). The presence of prestigious exotic goods in burials suggests that an élite had risen as more intensive productive modes were adopted. They would have controlled specialists' activities, such as the making of prestige pots (in a manner comparable with the Fournoi obsidian production centre of the Southern Argolid), as well as the "distribution and/or consumption of resources" (Barker 1995, 128-129).

Barker relies almost exclusively for data from other parts of Italy to present a picture of the earlier and mid 3rd millennium BC. However, the survey produced more direct evidence for the Bronze Age Apennine and SubApennine phases. The location of mid and low valley sites, Fonte Maggio and Masseria Mammarella, in close proximity to water with access to free draining soils, in conjunction with environmental data, suggest that the pattern of mixed agriculture which prevailed in Italy generally was common to the Biferno Valley (Barker 1995, 151). In the high valley there was a lack of permanent settlement, although pastoral and hunting activities took place high in the mountains.

With few examples to hand, Barker argues that single family settlements were unlikely because of the cooperation needed to farm effectively. This is somewhat in the face of ^{14}C evidence that the multiple dwellings at Fonte Maggio (Barker 1995a, 35) were unlikely to have been contemporary with each other (Barker 1995, 157); it appears, also, to assume that the isolated family farmstead had been left behind in the Early Neolithic, and had no further part to play in socio-economic evolution.

The Bronze Age landscape remained extensively closed and forested, with settlements situated in infrequent clearings, a pattern which changes gradually through the Iron Age. During the Iron Age, in the lower valley pottery reflects contact with Magna Graecia, but from the mid valley upwards styles changed little over the period, suggesting relative isolation. Domestic sites of the period are represented by surface scatters which vary in size by as much as a factor of ten, with by far the greatest number in the middle and lower valley (Barker 1995, 162). The livestock regime seems broadly similar to that for the Bronze Age, but the data were very limited. Crops included emmer, barley and beans and possibly grapes; the possibility that the latter may have been cultivated for wine production is suggested by fine wares at larger, high-status settlements (Barker 1995, 168-171).

Graves in neighbouring areas are indicative of the development of distinctly hierarchical, warrior societies with authority conveyed through kinship. It is argued that in the Biferno Valley such a social structure was beginning to emerge, but that the lack of high status infant burials from three Iron Age cemeteries in the upper and lower valley, where goods do show differentiation amongst adults (Barker 1995, 171-172), demonstrates that "status was achieved in life [not] ascribed by birth" (Barker 1995, 179).

It is only with the emergence of the Samnites that a more complex system of authority becomes established, in the second half of the 1st millennium BC. The early part of the period is characterised by the development of major hillforts and sanctuaries in the upper valley region of the Samnite Pentri, whilst in the lower valley there are the beginnings of urbanisation. By the 3rd and 2nd centuries BC the settlement hierarchy ranges from the lower valley city of Larinum and the upper valley centres at Boiano, Monte Vairano and Sepino, through larger nucleated village settlement, to villa/estates, and to humble farmsteads (Barker 1995, 208-209) set in a landscape undergoing massive clearance (Barker 1995, 313; Table 14). There seems to have been organised production of textiles, perhaps based on an outworker system. The plentiful examples of prestigious exotic items which might well have been imported. The organising of production, export and consumption is unlikely to have been restricted to an individual chief even at local level, and more complex administrative systems must have been developed.

In the opening two decades of the 1st century BC the volatile political situation in Rome had an increasingly unsettling effect on the valley, and culminated in the historically attested ravages of Sulla. Monte Vairano may have been deliberately destroyed (Barker 1995, 218) and most other hillforts seem to have been abandoned within the next few decades.

During this period some shrines, which the Samnites had built, and at which they had deposited prestigious items, seem to have been maintained, but only by the poorer members of the indigenous population. Before Romanisation the shrines' ideological status was on a par with that of the administrative centres; now they became peripheral to a centrally-based administration and state religion presented in monumental form in urban settings (Barker 1995, 242). It would appear that religion had been subsumed within the apparatus of the state.

The attraction of the centre brought about a decline in peasant farm-holding as people were drawn to the burgeoning towns, while the wealthier Samnite-descended families aspired to political influence in the greatest centre of all, Rome. Individuals could only hope to compete for such position through the acquisition of wealth, and although towns may have benefited by the display of that wealth in public works "absentee senatorial landlords... almost certainly took more from the valley than they gave back" through the surplus generated by slave labour on their estates (Barker 1995, 250). By the Early Imperial period the rural dispossessed had parented children, many of whom had to be fed through "alimentary schemes". In the lower valley, the land they had abandoned enabled bailiffs, and/or tenants/freeholders to farm tracts of land, although often, on behalf of the rich landowners in Larinum (Barker 1995, 233-234). By the latter 2nd and early 3rd century AD there were very few small farmsteads or agricultural buildings, but there is no evidence that this was a symptom of a continued population shift towards the nucleated settlements, which also seem to have been in decline (Barker 1995, 225).

During this period old trading links with the East Mediterranean had been broken and new ones forged with North Africa, and it is the increased incidence of fine wares from that area in the 4th century, which signifies a degree of economic revival in the middle and lower valley, where soils were better, and seaborne trade routes were easily accessible. However, investment in the urban centres of the upper valley failed to bring about prosperity there, and by the following century, prey to Visigothic campaigns, the economy of the whole valley went into decline again. Much of central southern Italy was excused payment of taxes at this time, and people from the area were still struggling to pay taxes by the middle of the century. Simultaneously, there was a marked decline in imports, and the results of the survey suggest a general decline in population (Barker 1995, 242; 250).

The story is continued from the sporadic settlement of the 6th century AD, through the development of medieval hilltop villages, upto the present. Barker has written that he had approached the valley infused with a view of a timeless peasantry grinding out a harsh living in an unforgiving, unchanging environment, generally overlooked by the rest of the world. What he found was very different. The grind of existence had persisted, but the population had moved to and from the countryside; in the past century tens of thousands of peasants have emigrated to America. The valley itself, which without human intervention would have been largely forested, retains some coastal pines and patches of deciduous woodland in the valley, but is otherwise an eroded landscape, scarred by Samnite/Roman and modern agriculture (Barker 1995, 66-67). Large volumes of soil have been redeposited and in places silts formed since the Early Medieval period have reached depths of 7m.

2.5v   Danebury Environs Programme

The Danebury Environs Programme is quite different from all other surveys treated in this book. Eschewing surface or ploughzone techniques in favour of geophysical survey and excavation at eight foci (Woolbury and Stockbridge Down; Bury Hill, Upper Clatford; Suddern Farm, Middle Wallop; New Buildings and Fiveways, Longstock; Nettlebank Copse, Wherwell; Houghton Down, Stockbridge; Windy Dido), it only has claims to being continuous survey because the fieldwork tests hypothetical chronological base maps derived from Rog Palmer's interpretation of air photographs taken over 60 years (Palmer 1984), covering an area of 450 sq km. Palmer's interpretations were partly informed by the extensive good quality, well recorded, fieldwork carried out in many parts of the region during much of the 20th century, and Barry Cunliffe has made extensive use of this literature. To accommodate the mass of data garnered by intensive means a multilayered publication strategy has been employed. Data and analysis from each individual site has been presented as a separately bound part of a volume two (Cunliffe and Poole 2000a, b, c, d, e, f and g, with respect to the list of sites above), with a synthesis in volume one. Volume one summarises previous research, the programme's strategy, old and new documentary evidence (including air photography and geophysical survey) and presents ecofact and artefact studies, including the all important ceramic series. There follow period by period thematic accounts, followed by a final broad brush summary headed the *longue dureé* (Cunliffe 2000). In effect the layering has the virtue of displaying a transparent synthetic spiral of inferential speculation.

The solid geology within the study area largely comprises Upper Chalk, broken only by Tertiary clays to the extreme south. The roughly north northeast to south southwest drainage of the rivers Bourne and Test provide natural definition of the west and east of the area, with tributaries to the latter enclosing Danebury on three sides at distances varying from 2 to 8 km. Recent water meadow management has altered the Test from a single main to a braided course and it is assumed that the modern springline is substantially lower than that of the target periods. The soils range from depleted, very tractable rendzinas on the chalk, through heavier paleo-argillic brown earths over the clay-with-flints, often wooded, to gravels and sands on the fringes of the Test valley, and alluvium. It is presumed that, allowing for some variation over time in the distribution of these resources, prehistoric agricultural potential would include cultivation, grazing of animals, pannage and fodder, coppicing and timber acquisition (Cunliffe 2000, 136-141).

Reconstruction of the cognitive environment of the area's prehistoric inhabitants was an important narrative objective for the programme, although Cunliffe had very modest expectations of what might be achieved. Preceding the attempt with a generalised characterisation of the topography, soils and hydrology of the area (Cunliffe 2000, 140-142), sucessive accounts are given for the periods within a range from 7500 BC to AD 50. Here I summarise the narrative for the latter part of the range, from around 1500 BC onwards, since this is the earliest of four periods targeted by the excavation programme. The first period is defined by a ceramic-based relative chronology, the succeeding periods by detailed ceramic chronologies tied to suites of carbon dates, rendering this the tightest sequence in Britain for the 1st millennium BC.

Cunliffe describes 1500-800 BC as a period of transitions. Where previously the population had created reference *points* in a mainly open landscape (monuments such as barrows), this was a period when the landscape was increasingly determined. It spans the currency of Deverel-Rimbury, post-Deverel-Rimbury and Early All Cannings Cross ceramics which, given the lack of sufficient cogent evidence, Cunliffe prefers to compress into *Classical Deverel-Rimbury* (ca. 1500-1100 BC) and *Late Deverel-Rimbury* phases (ca. 1100-800 BC. Cunliffe 2000, 149-150). Sometime around the mid 2nd millennium BC small settlement enclosures were defined, but not defended, by slight ditches are constructed, perhaps at a time when the first coaxial field systems were being laid out, respecting or sometimes utilising as markers, standing barrows. The regularity of the field systems, some covering as much as 90 ha, suggests that they are single concept units marking a major shift from dwelling in a landscape to taking possession of land, exhibiting "a perceived sense of order backed by the coercive power needed to implement such schemes" (Cunliffe 2000, 161), presumably not at odds with the view that as parcelled land they represent "a statement of communal control" (Cunliffe 2000, 194). The most substantial long linear boundaries are often later than these systems, cutting across them, apparently putting at least some components out of commission. In contrast, others are integrated into hilltop enclosures, as at Danebury itself, and in some cases continue in use into the Iron Age. Precise phasing was difficult because of multiple, comprehensive, re-cutting of substantial lengths of ditches from at least the Late Bronze Age onwards (Cunliffe 2000, 155-160).

Treatment of the dead varies over time and space. In the mid 2nd millennium BC inhumation associated with round barrows was still practiced, although cremations were already a feature. Indeed, at Kimpton the earliest phase of burial is from around 2000 BC, with continuity into the latest Bronze Age or even the Early Iron Age (Cunliffe 2000, 161). Elsewhere there is a marked decline in human remains, although the discovery of three apparently casually placed complete skeletons and a foot where chalk had been quarried from a ditch at New Buildings may indicate a connection with fertility. Cunliffe suggests that at least one use for the chalk would have been marling, breaking down and making productive the heavier soils. The dead are given up to fertilise the ground from which the chalk came as propitiation to the chthonic deities. The rite of giving up is represented also by deposits of metalwork in water and the earth.

The phasing and dating of the latest part of this and the earlier part of the next period, 800-300 BC, carries with it the sense of a specialist looking over the shoulder. Reports of nearby Late Bronze Age / Early Iron Age middens, Potterne (Morris in Lawson 2000, 166) and East Chisenbury (McOmish 1994 and 1996) and their well stratified ceramic sequences show refinements in the sequence. Present understanding is of a range of five distinct typological groups, each introduced and extensively overlapping during the currency of earlier introductions (Cunliffe 2000, 163). At the beginning of the period the enclosed hilltops at Danebury and Balksbury are major presences in the area, but the latter falls into disuse. There are traces of continuing activity on Danebury which, along with Bury Hill (a kilometre south of, and across the river Anna from) Balksbury, is defined by a new rampart and ditch shortly before 500 BC.

Within a century there is a dramatic burgeoning of new hillforts at Quarley (like Danebury, integrated into the scheme of major long linears), Figsbury and Woolbury. However, neither Woolbury nor Bury Hill show signs of occupation, and activity at Figsbury and Quarley appears "sporadic and short-lived". Cunliffe notes that all of these enclosures have opposing entrances, with the exception of Bury Hill where later Iron Age activity has concealed the evidence. He sees these as places for seasonal assembly and ritual. Danebury is distinguished from the others by storage pits clustering around possible rectangular shrines in its centre, and circular houses in the lee of the ramparts.

Elsewhere settlement comprises mainly enclosed an unenclosed "farmsteads", smaller "farms", all probably providing for extended family units, and one enclosed nucleated settlement, indicated by both greater density of structures and greater extent. Settlements appear to occur no more frequently than every kilometer, often on the "upper flanks of low hills" (Cunliffe 2000, 162-170).

An enclosure at New Buildings had a particular, perhaps ancillary, relationship to Danebury, to which it was linked by a double-ditched linear, but there are signs that most of the farmsteads interacted in a specific way with the central place. Crops, including spelt and six-row hulled barley, were threshed and winnowed at the farmsteads, then transported to Danebury for bulk storage, perhaps as seed corn for the following year. All the farmsteads are associated with animal husbandry, dominated by sheep, then cattle and, much less frequently, pigs. Animal by-products would have been important parts of the diet, whilst also supplying the raw materials for textiles and some tools. There is evidence for winter culling of older animals (the potential of cattle for draught may have contributed to their longevity) and a higher percentage of cattle at two farmsteads may be an indication of status variation between the smaller settlements. Seasonal variation in mortality shows further variation. The sheep appear to have moved well away from some sites during certain parts of the year, and the lack of dead lambs at one site suggests that lambing did not take place there.

Cunliffe draws a picture of an Early Iron Age belief system in which everything issues from and, ultimately, returns to the earth. The dead nourish areas where chalk has been removed to improve the productivity of the soil, and the massive middens associated with Late Bronze Age ritual give way to the more intimate introduction of rubbish in pits which have been removed from the cycle of storage.

Political narrative is restricted to the view that this was a hierarchical society, with a coercive force enabling the construction of major earthworks. The character of force remains unspecified. Wider ceramic distributions suggest that Danebury east territorial boundary was beyond the river Test, and that the Quarley linear and Devil's Dyke may have marked its west and north extents, and woodland on the heavy soils to the south. In the wider world, the material culture provides evidence for a high level of interaction, direct or indirect, reaching as far west as the Somerset Levels, and south west to Dorset, but with direct exchange zone perhaps including Salisbury Plain and the south coast via the Avon Valley.

Cunliffe speculates that there may have been less intensive activity on Danebury hillfort in the period leading up to and around 300 BC. But in the second quarter of the 3rd century BC there is reversal of this state of affairs. The closure of its south west gate exercised a change in patterns of movement in, and at least immediately around the hillfort, at a time when it is evolving into its developed form. The sequence is once again traced through special attention to the ceramics, with the identification of overlapping phases 6 and 7. Within this framework a distinctive geographical distribution is noted, with the slightly earlier-originating Yarnbury-Highfield style occurring west of the Test, including a particular focus around the upper Avon and its tributaries (Cunliffe 2000, fig. 4.29) and the St. Catherine's Hill-Worthy Down style featuring to the east of the Test and Anna / Anton.

What strikes Cunliffe is the lack of known incidence of the latter style within a radius approaching 10 km around Danebury, excepting the hillfort itself, where it is dominant. The next two centuries bring the peak of Danebury's activity as a central place for redistribution of resources, settlement, crop storage, and ritual, apparently within a pattern of extreme nucleation. The "inner zone" would have been farmed by the hillfort's inhabitants, and an "outer zone" both by hillfort inhabitants and those populating farmsteads from further afield, the latter having only a loose and changeable relationship with the centre (Cunliffe 2000, 184-185. Fig. 4.31). Crucial parts of the productive cycle such as lambing and calving appear to have been managed very close to the hillforts, judging by the high incidence of neonatal remains. Towards the end of the period there was a reinvestment in Bury Hill, marked by the enclosure of a smaller area. Comparison of the bone assemblages shows that whilst horse comprised 4% of the total at Danebury, it was 48% of the total at Bury Hill, seemingly an indicator of variation in specialisation between sites, be that in the form of breeding the animals or ritual deposition of artefacts associated with them (Cunliffe 2000, 189 and 191).

Within the wider scheme the presence on Danebury of both staples (such as iron, salt, querns) and luxuries (bronze, shale and more exotic materials) demonstrate links to the north or west, and particularly to the south. The port of Hengistbury Head was busy during this period and 40% of the ceramic assemblage has been diagnosed as of the St. Catherine's Hill-Worthy Down style found at Danebury. This suggests that the hillfort exerted influence as far west as Salisbury Plain's River Avon (Cunliffe 2000, 182-183).

Ritual practice, restricted to observations from Danebury, seems little different to what went before. A marked increase in the amount of human remains is easily accounted for by the increase in population, although the volume of male remains is now matched by the number of female. Otherwise the deposition of iron artefacts under a roundhouse floor seems likely to represent some sort of offering to the gods, perhaps for protection of the household (Cunliffe 2000, 184).

Cunliffe identifies 100 BC to AD 50 as another period of transition. Dating is governed by ceramics with stylistic and technological affinities with the continent, described as Southern Atrebatic (Cunliffe 2000, 186). By around 100 BC Bury Hill has been reoccupied, just within the bounds of Danebury's hypothetical territory and possibly directly rivalling its authority; alternatively, it is suggested that the hierarchy at Danebury simply moved. The earlier part of the period is associated with the burning down of Danebury's remaining gate and there is a dramatic decrease in activity, thereafter. The presence of a large number of slingstones around the entrance at the time of its destruction suggests that defensive potential of the ramparts went beyond mere symbolism, and Cunliffe speculates that warfare may have been endemic at the time, perhaps prompted by growing competition for good land. The lack of precision of the ceramic dating renders it unclear whether Bury Hill's resurgence continued beyond this episode, although it is assumed so, but by 60 AD activity at both hills is sparse.

At around this time substantially enclosed areas of up to 3 ha become loci for settlement, contrasting with less structurally imposing clustered enclosure settlements, some of which were within the bounds of former hillforts. A banjo enclosure ditch was dug at Nettlebank Copse, another area which disappeared from the archaeological record during Danebury's zenith. Despite a lack of structural features there were animal skeletal and midden deposits associated with the enclosure which was integrated into the wider scheme of boundaries and enclosures by antennae ditches. It is suggested that it was an area of activity for the wider community, perhaps associated with the bringing together of herds and flocks, and with feasting.

In some areas cattle become more important, whilst sheep often lived longer, suggesting that their by-products, milk and wool, were utilised more frequently. New crops, oats and peas, supplemented the old, and were grown separately. But perhaps the most striking feature of the economy is the evidence for much increased exchange exemplified by ceramic assemblages from several settlements comprising 10 to 40% Poole Harbour wares. The trading of the commodities stored in these vessels must have become very routine. Amphora and new weeds amongst the harvested crops indicate increasing contact with the continent (Cunliffe 2000, 189).

During much of this period the dead virtually disappear from the record, and when they reappear in the 1st century AD it is often as cremated remains, although a continuing tradition of special deposition in pits shows that some old rites persisted. But the introduction of similar material into enclosure ditches at leasts suggests a change of emphasis since a sequence of deposits in such a ditch is likely to be over a much longer period of time.

The Danebury Environs Programme narrative has fine skeletal structure derived from Rog Palmer's air photographic interpretations. It is doubtful that any other survey targeting British prehistory can claim to have offered so detailed an account for areas of muscles, sinew and flesh. We have unparalleled detail of its central nervous system, too. The leading question is: how much can we infer about the rest of the body from a small part of the biceps here, or stomach wall there?

2.6 Discussion

The narratives outlined above reflect radically different approaches. The tight survey methodology of Foley elicits a compelling but spare, model-based, narrative rooted in an explicit account of prehistoric and modern ecology onto which the residues of human activity have been mapped. In contrast, Barrett et al.'s loud proclamation of the reproductive theme sometimes masks a particularly fine line between data and narrative. Taken at face value, however, the authors have elicited political accounts from material where others might have contented themselves with a fairly basic economic outline. The most fundamental contrast is in the method of information-gathering. The Amboseli project used previously published regional ethnographic, ecological, and climatic data, to support Foley's stratified random sampling. The Cranborne Chase project relishes the incorporation of a diverse and inconsistent literature derived largely from site excavation, and frequently integrates, perhaps with undue seamlessness, informal surface collection and accounts from beyond the periphery of the study area. Indeed, with the exception of molluscan and carbonised material from an extremely restricted area, the lack of local waterlogged deposits has compelled the authors to make inferences about ecological change from surveys outside the area. Woodward (1991) has gone further, producing maps which chart the decline over the millennia of woodland on the South Dorset Ridgeway, again from a combination of exotic data and mollusca from excavated features.

In the narrative expositions of surveys carried out in East Africa, Europe and North America climate has varied from the semi-arid to the temperate and the topography from the valley bottom to high on mountain-sides. The temporal focus has been prehistorical. With that in mind the inclusion of class G narratives may be seen as discretionary. It is notable that the most empirically framed surveys, those in the Amboseli, Owens Valley and the Great Basin were conducted with least inclination to pursue the elusive issue of power. Each offers insights into economic/subsistence behaviour, but only in the case of Thy, where site functions vary and labour must surely have been divided, do we observe the potential for social tensions which might lead to the prioritising of behaviour in turn bringing competition for authority embodied in particular individuals or groups.

It is surprising that the issue of power has not been raised within the evolutionary perspective of Larson et al., although a fuller discussion would have been prohibited within the journal format. Even so, one of the stabilising strategies successfully predicted from within the paradigm was trade, implying organisation and generation of surplus for exchange and, consequently, inter-group negotiation.

Still more surprising is the veneer which passes for ideological and social class G narrative in the Dalmatia project. The meta-models, in particular that of Arenas of Social Power, aspire to presenting data in the light of the inherently dynamic concept of social reproduction. Generalisations such as that kinship networks being maintained for "the cooperative pooling of labour" have the appearance of being borrowed from elsewhere, rather than having an organic place within a well constructed argument. Equally, insights suggesting that the annual clearance of stones eventually came to have regenerative connotations by symbolising harvest need well argued foundations if they are to be seen as anything other than absurd. Chapman et al. have had more success at ranking the soils than at understanding the patterns of authority in human societies.

In contrast, Woodward has set out a compelling account of the development of authority through symbolic and material power from the Neolithic and through the Early Bronze Age in Dorset, despite omitting to set out a particular paradigm. As with his influential earlier work he employs a problem solving strategy; but whereas he went little further than class E distributional interpretation in the Great Ouse Valley (Woodward 1978, 50), in the case of the Ridgeway he has borrowed landscape concepts to invest a chronological sequence with a persuasive human and environmental dynamic narrative. The images, both in words and pictures, are vivid; but some of the cogency may well reside in the story-telling itself, as much as in the quality of the data.

There is a fine line between making the most of data analysis, and discarding data as an encumbrance; John Barrett, in the company of Hodder, Shanks, Tilley and others, stands accused of the latter. It is fortunate that he has the skill and desire to explain his perspective. At the heart of it is the human-as-agent, a biological being transformed "into a sociable being through knowledgeable action", so necessarily including consciousness (Barrett 1994, 165). If this distillation of what it is to be human is accepted, then a leap of the kind made by Hegel (Hegel 1977) is inevitable: although I do not actually recognise myself in that other (the "agent" who at this moment is the object of my research), I recognise features of his/her agency which are analogous with my own being. In short:

> "if the different emphases which we employ to understand others are taken to have a general application in understanding all humanity (which, for them to operate, they must) then they will also be the principles which we will employ in understanding ourselves, and they will also inform our own attitudes towards the validation of archaeological knowledge." (Barrett 1994, 165)

This is not a fresh insight; nor, indeed is the recognition of grammars of experience (Barrett 1994, 167); the analogies we perceive in our ancestor's lives can only be formed because we infer characteristics of behaviour in the testimonies we choose from the artefacts we confront, be they tools or modified landscapes. Into that confrontation, which is merely one aspect of the routine processes of understanding and acting by which we survive each day, we bring qualitative perceptions, such as "sharp", "durable", "serrated", "pleasing", "useful"; these are words of our now, but they bear the resonances, and are structured in patterns, which evolve and have co-existed over millennia. In this aspect of recognising mutuality in consciousness (albeit, former consciousness in the case of our object) we engage in a moment of archaeological experience.

We must reconcile ourselves to working within an explicit understanding of there being a plurality of modes of consciousness for ourselves and our objects, or we must choose to work within a fundamentally untrue framework, with our object appearing static within a system shaped by external laws. Static models have served, and will continue to serve, slices for our experiences of the past; but the slices compress synchronic and diachronic consciousnesses within which reside the decision-making that shaped material culture and discard behaviour. A falsely homogenous "society" is created.

The approach which Barrett and others seek to legitimate, is one where the archaeologist places him/ herself within the discursive framework of the object. To take a place in that discourse the archaeologist must be able to recreate the most intimate and the largest scale aspects of the object's environment and then place him/herself into that environment. From here the archaeologist can explore the relationship between the diversity of recovered and inferred artefacts, their relationship to the landscape, and to the agent.

Even if this is a legitimate aspiration, is it feasible? Tilley, too (Tilley 1994, 170-201), has considered the Dorset Cursus from the landscape perspective, but chooses to emphasise linear motion within it as the means for grasping its meaning. Whilst acknowledging the aptness of Bradley's view that the Cursus is to be understood from within he goes further, asserting that only by entering between it north east terminals and walking the full 10km to its south west end, can we appreciate the moments of surprise and revelation which endow the long barrows within and outside it with their particular significance. Through a juxtaposed sequence of text, figures and photographs his account of a physical journey through a specific landscape offers a coherence with which, in general terms at least, it is hard to disagree. However, it is only when there is sufficient evidence for the imaginative reconstruction of an ancient landscape that Tilley's approach is possible; thus, a walk around Stonehenge or Avebury, or the struggle up a track between the ramparts of a hillfort, informed by knowledge of the local landscape, may enhance our appreciation of a large scale feature. But walking where the timber structures of an Iron Age settlement once stood will achieve little unless we can find convincing means for the interpretation of fine detail.

Both Barrett et al. and Tilley offer class G ideological narratives. Aspects of ritual are characterised, and interpreted in a manner which gives coherent accounts of the evolution and overlapping of beliefs. The power of these accounts derive from their roots in intimate experience; the archaeologist places the reader in the presence of possible activities, ensuring that he/she recognises that they were enacted in a manner specific to a time, place and hypothetical agent.

Adopting a more empirically based perspective, the North Keos survey presented social and economic development through time, set within the limits of residual material culture, in the form of site and off-site scatters. Aside from the initial description of modern Keos, the presentation of environmental data was restricted to evidence or produce from period to period and the effects of geomorphology on surface collection. One of the great advantages of considering the evolution of the landscape is that on the one hand it brings to mind a more vivid, situated picture of human activity and on the other hand it can be used to cross-examine and inform interpretation based on artefact distributions.

It is precisely the landscape which the Southern Argolid survey has attempted to emphasise to allow the reader to appreciate fully the histories it describes. The flux of the landscape is brought to life when described separately from man but it becomes a passive backdrop to social and economic change once the human story is told, despite the promising title of chapter 6: "Fifty Thousand Years of Coevolution of Landscape and Human Settlement" (Jameson et al. 1994, 325-414). Humans dictate to the landscape in settlement and agricultural terms but, although they may choose a particular topography or soil for their activities, we see little of the reconceiving of succeeding generations relationships with the land that plays so important a role in "landscape archaeology".

Further criticism of the book has focused upon its "formalist thinking" (Alcock & Cherry 1996, 210). The authors have looked for common patterns of economic behaviour, in particular external trade. The immediate consequence is that attention is drawn towards the elite who controlled production and exchange. Prosperity is thus a class-specific matter; the perspectives of the impoverished workers on the estate of an affluent owner go unconsidered. A similar complaint might be levelled at Barry Cunliffe's work in and around Danebury. Although the upper echelon of society is never characterised, the moving and shaking in the study area is generally attributed to coercive forces. In any event, although there is an expressed enthusiasm for the "cognitive landscape", we are presented with essentially a layered positivist account, with a responsible acknowledgement of alternative explanations. Within this framework there are bold hypotheses (for instance, the extreme nucleation at Danebury during the Middle Iron Age) which might be tested if resources permit!

In the work of John Barrett one senses a passion about how to view a landscape whereas an author such as Graeme Barker's passion is for the landscape and its peoples, almost to the point of idealism. Consequently his narratives are conveyed with great conviction. Although providing the sinews of the account, theory never obstructs the reader's view of the story in the way that it does in Chapman et al.'s account of work in Dalmatia. Samples from Barker's account of the Biferno Valley are analysed in *4.4iic*.

The theme of this chapter may be summarised as an assessment of narrative coherence, rather than narrative validity; in presenting synopses I have made passing references to methods of data retrieval and have paid some attention to the issues of modelling. Models can be introduced into a project at any stage before publication. The current fashion demands that a clear theoretical framework ought to be established at the outset, prior to fieldwork. However, theory frequently lags behind practice not only on the very general scale where techniques outstrip our understanding of their application potential, but also within individual projects. Models are frequently introduced into a project during writing-up when, by post hoc manipulation, they can provide foundations for a particular interpretation. The relationship between model and narrative can be dangerously close; nonetheless, an account must be answerable ultimately to the quality of its data.

One means for investigating the relationship between evidence and narrative is the introduction of new categories of evidence, such as contemporary texts. Only rarely will this give insights into the particular which can be generalised, but it can provide a valuable context or elliptical insights into residues of past behaviour.

3: Acts of Regional Survey

3.1 Historical summary of survey approaches

The tradition of recording relics, monuments and documents in particular locales has a long history. In Britain there were William of Malmesbury's (Radford 1968) scholarly attempts to link institutions in particular places with events during the 12th century, and a lineage of peripatetic antiquarians continues through the likes of John Leland (16th century) and William Stukely (18th century). By the 19th century European colonial expansion, coupled with the rationalist spirit of information collating and categorisation, encouraged the pursuit of natural scientific and historical studies. In the case of the latter there was an acknowledged ideological goal: the recapturing of physical links with a classical past concealed within the Turkish empire. The recording of structural remains in North Africa, the Middle East and the northern Mediterranean paved the way for campaigns of excavation by Schliemann, Woolley, Flinders Petrie and many others from the late 19th century onwards. By this time survey in the shape of the graphical recording of structural remains was a routine event for colonial forces, carried out by skilled military and civil service draughts people.

Ethnographic observations by J.H. Steward (Bradfield 1973, 349ff.) from the 1930s onwards, in the Great Basin of the United States, provided one of the first theoretically conceived models for human behaviour within a regional framework. But in Europe "classical" surveys carried out after the second world war had more in common with the practices of the colonial antiquarian than with any rigorous attempt to reconstruct full patterns of settlement land use.

While mesoamerican studies benefited from the positivist "middle-range theory" models offered by Binford from the 1960s onwards, it was only at the end of that decade, following the publication of C. Vita Finzi's *The Mediterranean Valleys: Geological Changes in Historical Times* (1969) that the Europeans finally accepted the value of detailed and wide-ranging surveys, using data from complimentary disciplines. Although Ward-Perkins work from the British School at Rome had already started to reveal unsuspected settlement patterns in Italy, it was not until the very impressive results from the Melos (Renfrew & Wagstaff 1982) and Boeotia (Bintliff & Snodgrass 1985) surveys entered the literature that systematic regional field methods were shown to be worth the level of investment which they required, although no publisher has been found for a full finds report from the former, and the latter has yet to be made available as a monograph.

The demise of haphazard surface collection and recording is aptly exemplified in a vicious debate between John Cherry and Richard Hope Simpson, which took place in the early 1980s (Cherry 1983; Hope Simpson 1983). Already, though, the more important topic was the comparative efficacy of extensive against intensive survey, an issue which the weight of practice seems to be resolving in favour of the latter. Since then new choices have confronted project coordinators: a wealth of techniques have become available through scientific developments enabling not only the collection of new types of data, but also radical and more widely available modes of presentation, using computers. Major surveys such as that of the Roman Wroxeter, Shropshire (White, R; van Leusen, PM) have used magnetic, electrical and ground penetrating radar surveys to build up a pseudo 3-dimensional picture of the city; the hinterland has been subject to surface collection and, combined with geological and climatic information, it has been represented graphically using a geographical information systems (GIS) programme.

Useful summaries of regional survey programmes since 1980 can be found in Keller and Rupp (1983), Macready and Thompson (1985), Barker and Lloyd (1991), Mattingly (1992) and Alcock (1993). More upto date information can be gained from the University of Michigan "Home Page" on the Worldwide Web.

3.2 Determining the land

The notion of "regional" in regional survey is arbitrarily variable. It can encompass several thousand square kilometres, or only a dozen or so. "Region" is not a determination of scale, rather it is one of approach. I would argue that a survey which employs detailed topographical investigation, supported by augering over an area of a few hectares such as that at Miwa, Japan (Barnes 1993) is regional in perspective, albeit at the extremely micro level. Although excavation took place, and fieldwalking did not, the idea of "site" was dissolved in favour of research into a particular landscape.

How may the "region" be distinguished from the "site"? Meaningful distinction can only be achieved on a plane of conceptual comparability: what both concepts have in common is that they are spatial units which, characteristically, are subdivided. An excavation site may comprise a series of trenches of variable size, in turn subdivided either in measured units, or in the form of archaeological targets such as contexts and features. In essence the site comprises relative (archaeological) and arbitrary entities. Even if we accept that the artefact is the basic unit of archaeology, common to both regional and site perspectives (not always uncontroversial; for the majority of surveys before 1985 the "site" was considered to be the basic unit of regional survey), it remains essential for us to have a means for investing it with meaning greater than its material existence. In the first place that meaning is derived from its place within the set of basic units: its identification within a project's assemblage, and within the framework of comparable objects from elsewhere. Almost all other points of association are related to the object's position in physical space. This includes arbitrarily (geometrically) defined space, relative (archaeological: "contexts", "features", "landscapes") space, and correspondence with other objects, some of which will fall within Clarke's definition of an artefact (to include such ecofacts as burnt residues, harvested grain, bone of domesticated or hunted animals etc.), some of which may be natural components, for instance soil, which provides a geomorphological context for an object. Of course, many soils conform to the definition of artefact; very frequently geomorphology is a by-product of erosion, induced by human activity.

The construction of meaning for the artefact recovered during regional survey is the same, in principle. Apart from its typological associations the artefact must be contextualised in arbitrary and relative (archaeological) space. At a large scale it belongs within a region or survey area, although it also has a place within a macro-region once issues concerning raw material sources and exchange are taken into account. The region may be considered to have archaeological integrity, although in most cases it is an area designated arbitrarily. Some archaeologists might argue that in this there is fundamental divergence from the notion of site, but the briefest consideration will show that notions of definition and limit are highly subjective. While it is perfectly acceptable on a pragmatic basis to refer to the hillfort of "Maiden Castle", it very quickly becomes problematical to specify exactly where the hillfort ends. In the first instance it has changed shape and size, especially during the period when it assumed most of the characteristics by which it is recognised; but in any event, excavation on the periphery or outside a gate, or alongside the external rampart ditch, is very likely to reveal contexts which were created through activity which is not only contemporary with the Iron Age use of the hill, but which is part of the activity of the hill. This is not to deny the legitimacy of the notion of the hillfort, "Maiden Castle": it is simply a statement that investigations of that hill, and its environs, cannot give a theoretically sustainable account of its limits as an archaeological *site*.

Which areal units do have archaeological integrity? In essence they are definably discrete items. Although a context may consist of several components, they form a sufficiently homogenous whole by which the archaeologist may distinguish them from another multi-component unit. The equivalent in the region is the stratum - a mappable set of attributes which may be geographical, topographical, ecological etc. All other spatial units of excavation or regional survey are arbitrary. Modern regional survey uses a wide range of techniques; their efficacy can only be measured in spatial terms. Unfortunately, the conceptualisation of archaeological and arbitrary space varies from one survey to the next, so before outlining field techniques and methods it is necessary to define a set of terms. Most of the terms can be found in existing survey publications, but they often vary in meaning. Although no one set of terms can be imposed retrospectively, or on future projects, the definitions given below aim to provide a common vocabulary which I shall adopt in this study, regardless of that adopted elsewhere.

1) **Survey Area** The fullest extent of the area designated for research, within which sampling procedures will take place.

2) **Stratum** A sub-area for sampling distinguished from other sub-areas within a particular criteria range, such as ecological, topographical, or geological. Thus a set of categories for topographical strata within a single survey area might be "plateau", "ridge", "upper slope", "lower slope", "valley bottom", "plain", "coastal". Usually each stratum will be sampled with the aim of obtaining a statistically viable assessment of the stratum-specificity of the intensity and nature of human activity in the survey area.

3) **Transect** A pre-defined broad linear unit, within which sampling will take place, imposed upon the survey area. It is increasingly common for a transect to form a cross section of several strata.

4) **Segment** A pre-defined measured unit imposed either in a continuous mosaic over the whole survey area, or over selected portions of it. It is a means for ensuring more even distribution of collection units, typically for bias-reduction in stratified random sampling.

5) **Collection mosaic** A contiguous group of collection units.

6) **Collection unit** Either one unit from a collection mosaic of uniform pre-defined, map-grid determined locations, or one of an aggregation of non-uniform, distinctive landscape features (i.e. fields) of roughly similar area. A collection unit may comprise one or more meshes, grids or lines, but at the analytical stage it will be the smallest areal unit for statistical representation and comparison. Ideally, it will be the smallest area from which a statistically viable sample size can be obtained.

7) **Mesh** A pre-determined, fixed or randomly, selected pattern of lines or grids. Either a single mesh, or a pre-determined, consistent, number of meshes may form a single collection unit.

8) **Grid** A uniform, pre-defined square from which all artefacts, within the project's specified range of interest, are collected. One grid may form a single collection unit or, for pragmatic reasons during fieldwork, it may be one of several in a sampling mesh.

9) **Line** A uniform, pre-defined linear distance of measured or estimated width (usually between 1 and 5m) from which all artefacts, within the project's specified range of interest, are collected. Either a single line or a group of (usually parallel) lines may form a single collection unit.

10) **Sample point** The map reference chosen for the representation of a collection unit (or units), which will be used for the graphical and statistical presentation of data either in absolute figures, or as an area ratio. Within a GIS framework this might become a single rasta pixel, and the base for the creation of a polygon.

11) **Area Ratio** The expression of the rate of artefacts collected or observed per given measured area. The dimensions of the area are independent from those of collection unit, grids or lines. Ideally its dimensions are determined by the smallest size of area (i.e. per ha, 1m^2 etc.) which yields meaningful statistical analysis.

Table 3.1. Terms of spatial definition

This list includes terms appropriate to fixed, repeatable, measured spatial units, and to spatial units defined by their particular "natural" characteristics. Although I have, for the most part, given terms which already exist in the literature, everyone of them has at least one synonym. I have constructed the list in an order from largest to smallest area, but note that no one survey would use all of them.

If included, elements 3-11 of this scheme are necessarily present from the planning stage, through fieldwork, to analysis. They may be discarded from the interpretative/ narrative phase in favour of the fruits of the analytical process. Contrastingly, strata do not have to be introduced until the analytical stage; indeed, the elements by which the strata are defined may only be decided from fieldwork data. In general, a project testing models/hypotheses is likely to use pre-defined strata (Carreté et al. 1995; Shennan 1985, 5; Foley 1981; Bettinger 1977; Barker 1995), in so far as it will set out to sample specific variables. The variable may be crude, such as "intermontane basin/upper valley", "middle valley" and "lower valley" (Barker 1995), or more specific, such as "Upper Chalk... Clay-with-Flints, Middle and Lower Chalk, Upper Greensand, Gault Clay and Lower Greensand" (Shennan 1985, 5), to the much more sophisticated ecological units used in the Amboseli (Foley 1981) and Owens Valley (Bettinger 1977). In the case of the Tarragona regional survey geological and topographical characteristics influenced the layout of the four transects, but from the fieldwalking it transpired that an additional topographic category, the low rise (Carreté et al. 1995, 245) was the prime determining factor for settlement location.

Nearly all modern archaeological survey will present, at the very least, a geological account; most will describe the distribution of modern vegetation and landuse, and the better surveys will make some attempt at reconstruction of past environments. However, this information only qualifies as strata if it is used specifically as a class of variable against which the distribution of artefacts is to be measured.

Although both the Dorset surveys (Barrett et al. 1991; Woodward 1991) provide adequate geological maps, neither meet the stratification criteria with respect to environmental reconstruction. They both give fairly full pictures of prehistoric vegetation, but the pollen samples came from outside the regions, while the molluscs retrieved from a very limited number of ditch sections cannot possibly carry the interpretative burden placed upon them (a risk noted by Gaffney & Tingle 1989, 4). Evans and Rouse (in Woodward 1991, mf5: A3) report that 200 auger borings divided between 7 transects (some with 25mm, some with 100mm bits) were of limited use; they failed to penetrate hard layers and missed thin strata. 15 test pits divided between the same transects were more informative, providing some useful environmental data, particularly mollusca, but a long trench in one transect provided far more detail, an approach vindicated elsewhere (Bell 1983). To rely principally on environmental data from excavated features risks circularity of argument such as: there must have been a clearance phase; there are woodland mollusc species in the features; therefore, the landscape was generally one of regenerating woodland. The Biferno survey is equally prone to make assumptions about the landscape (Barker 1995), although the more widely spread systematic fieldwork partly obviates the problem.

By using transects cutting across the grain of known geology, topography, or paleoenvironment a survey is better placed to make inferences about the distribution of environmental characteristics and of artefacts. Good use of strata occurs in the Neo-Thermal Dalmatia Project (Chapman et al. 1996, 27, fig. 13; 39-41), where parallel transects are laid out against the grain of various soil types, and the record sheets filled in during walking allowed fairly fine environmental discrimination between collection units, as well as enabling finds distribution mapping (Chapman et al. 1996, 47-51); a similar approach was used in the Tarragona region (Carreté et al. 1995, 46-49), and in Berkshire (Gaffney & Tingle 1989, 9, 32-33; Tingle 1991, 21).

Mattingly has offered an 8 point "evolution of survey methodology" (Mattingly 1992, 90) but it is misleading to place the development of surveys within a simply "progressive" framework. He drew attention to the intensifying of coverage and to increasing use of paleoenvironmental and geomorphological data. While these options are generally preferred now, it should be noted that in the 1970s both the Amboseli and Biferno Valley surveys placed great stress on geomorphology but, by comparison, the Tarragona hinterland project made much less use of such information during a programme of work from 1985-1990. Budgets still play a crucial role in what questions are asked during a survey, hence influencing the choice of techniques. Equally, researchers may feel that lacunae need filling within a particular academic paradigm (Gaffney & Tingle 1989, 7).

There are exceptions to any of these methodological patterns, notably the Dorset surveys which, in effect have compressed landscape and area units into just two tiers. More common in modern surveys are three and four-tiered areal component matrices:

a) survey area — segments— collection units: North Keos (Cherry et al. 1991)
 / Boeotia (Bintliff & Snodgrass 1985) / Nemea (Wright et al. 1990);

b) survey area — strata — segments — collection units: Amboseli Park (Foley 1981) / Owens Valley (Bettinger 1977);

c) survey area — strata — transects — collection units: Dalmatia (Chapman et al. 1996) / Tarragona region (Carreté et al. 1995) / Maddle Farm (Gaffney & Tingle 1989) / Vale of the White Horse (Tingle 1991).

The crucial differences between the three matrices lie in the use made of strata. At first sight Bettinger and Foley appear to be working in a similar fashion, but for the reasons discussed above they are working in quite different ways. The former required a good sample size to achieve detailed analysis of the assemblage composition of surface concentrations, with the expressed objective of detecting stratum-specific sets of tools which would demonstrate ecologically determined patterns of subsistence behaviour. Foley's small sample size per collection unit, coupled with the lack of chronological resolution, made the reconstruction of tool kits impossible. Instead, he assumed certain modes of behaviour to have taken place, and located them almost entirely on the basis of the volume of discarded material, rather than the collective functional characteristics of the artefacts.

Despite their different approaches, both Foley and Bettinger were able to contribute information about site hierarchy or, as the former would prefer, spatial components. In the Amboseli, deviations from the mean continuous distribution allow general discrimination between home range activities. Thus, no particular site is mapped, but the repeated use of an ecological niche for a certain range of activities may be postulated. Bettinger relied on particular constellations of artefacts to draw more general conclusions about the loci of activities; he was able to identify five ecologically-dependent settlement categories distributed across a 5,500 year time span. From a set of localised and specific examples derived from intensive coverage of random grids he made general inferences. The inclination to generalise has long been a topic of hot debate, and must be particularly susceptible to criticism when the method of survey, intra-stratigraphically random, makes no attempt to judge the uniformity of behaviour patterns across the whole of a particular ecological unit. In particular, it may fail to detect other categories of strata. Thus, if altitude and ecological characteristics are the definitive units, random sampling within them will not show topographical variation. In contrast, Foley's method of sampling within every segment made generalisation feasible.

The most commonly applied areal matrix is c, where collection units lie within broad transects. They offer a compromise between the overall representation but weakly resolved data of the segmented approach, and the potential bias of stratified random sampling.

Characteristically, the c matrix is employed by surveys to locate sites, and to discriminate between classes of sites within an economic system. They attempt to show regularities of scales and types of settlements on particular geologies, topographies etc., in order to make statements about the economic and political structures and identities of a region's inhabitants. Many modern surveys operate within the b matrix as a first stage; when higher artefact densities occur, or architectural material is observed, a second phase of intensive collection is implemented.

One of the most influential surveys from this group, indeed the only British survey which has had significant direct impact on project methods in Europe and elsewhere, was that conducted at Maddle Farm, Berkshire. The survey's expressed intention was to contribute a "study of the landscape and settlement context of a specific villa", rather than a more general examination of Romano-British settlement distribution on the chalk Downs (Gaffney & Tingle 1989, 7). To this end a set of four problems was addressed, which issued from the explicit assumption that there were politico-economic structural relationships between settlements. The initial stratification of the project was based on the geology, but after work within a 2km radius circular core area around Maddle Farm villa it was noted that the lie of the land was a determining factor in its use (Gaffney & Tingle 1989, 9). Therefore the second stage was stratified according to topography, as well as geology, in transects outside the core area in the villa locality.

These matrices offer a means of survey classification. They give an indication of a project's approach to survey, but fail to give indication of the spatial or chronological resolution aspired to or achieved. In themselves, they are not good indicators of the relationship between sampling strategies and their subsequent narratives, a theme which will be developed in chapter 5. What the matrices do provide is a framework within which techniques can be applied in the field, and much of the remainder of this chapter will be devoted to a description of those techniques and outlines of the information they may be expected to provide.

3.2i   How many fingers in the pie?

If it is assumed that pastry and sugar are ingredients common to all pies, how many times will Jack Horner have to put in his thumb to decide whether a particular example may be characterised as a plum or blackberry and apple pie - or even a combination of the two. Probably not many, and if his intention is to sample the fruitiest portions of pie his experience may suggest that he should put in his thumb where the undulations of the crust rise. On the other hand, as a gourmet, he may wish to judge the quality of the pie according to the evenness of the ingredients' distribution. Assuming that he wishes to share it with some friends, how many times must he insert his thumb, and is hole the size of his thumb going to be large enough to provide a representative sample of ingredients?

Most archaeologists would probably prefer to eat the whole pie then afterwards tell their friends what it was like. However, their pies are usually much too large be consumed at one meal, and there is a much greater variety and volume of ingredients, albeit unknown!

As survey has developed into a more rigorous discipline expectations of its potential have increased. In the 1970s and early 1980s some practitioners were clearly of the opinion that provided a sufficient portion of a region was sampled it would be acceptable to multiply up to achieve overall estimates of period by period settlement patterns. In recent years Martin Kuna (1990) has appeared willing to apply this assumption at a regional level, while Jon Steinberg (1996) has made detailed claims for the characters of individual sites within a region on a similar basis (see *4.3iiid*).

The debate over whether 2% or 10% of an area can offer a meaningful overall picture has moved on. It has been recognised that at least as important as the frequency of collection is the amount of material collected from each sample for each target period before a valid statistical analysis can be conducted. Consequently, more theoretical effort has been invested in finding appropriate means for covering the whole of study areas. In the case of fieldwalking variations between line intervals have been tested, and the subsequent trend has been for the distance to be reduced. Consequently the last variable, the size of the area of study has had to be reduced, unless a projects resources can be increased. These issues will be dealt with in more detail in the sections dealing with particular techniques outlined below, and in the assessment of survey comparability in chapter 5.

3.2ii   Etching the record: field and form

Portable artefacts form only one part of the information which may be collected. Much more can be recorded as it is encountered in the field. As a general rule forms will be required for noting the collection unit or site's grid reference, the landowner and other details of location. If more than

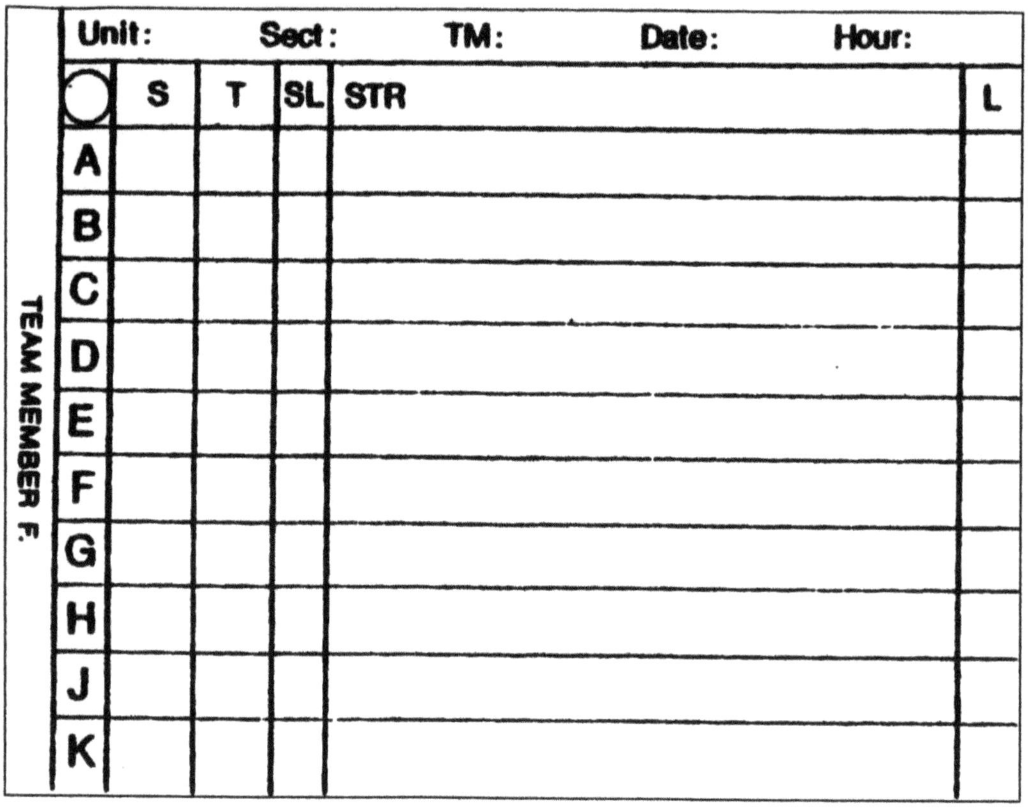

	S	T	SL	STR		L
A						
B						
C						
D						
E						
F						
G						
H						
J						
K						

Unit: Sect: TM: Date: Hour:

TEAM MEMBER F.

S = Number of pot sherds observed in 100m length. V = Number of tile sherds. SL = Number of pieces of slag. STR = Record of structures. L = Length in metres of incomplete lines

Figure 3.1. Team member record sheet, Africa Proconsularis (from Dietz et al. 1995, fig. 3)

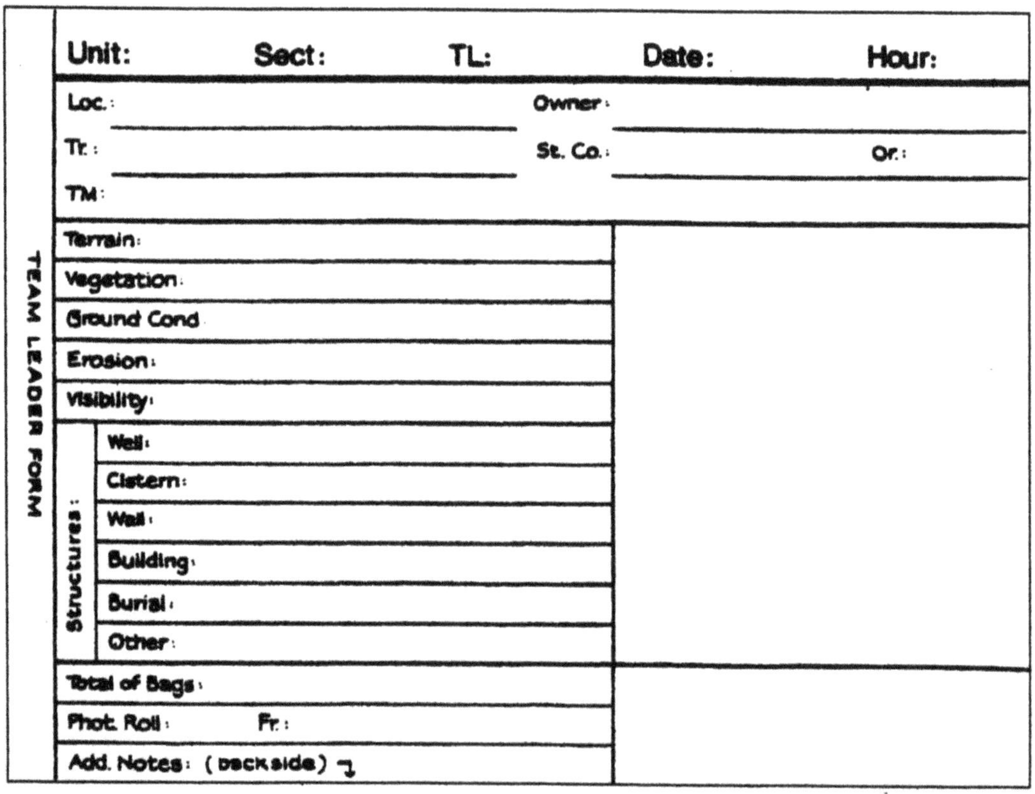

Figure 3.2. Team leader record sheet, Africa Proconsularis (from Dietz et al. 1995, fig. 4)

South Cadbury Environs Project

Fieldwalking: Line record sheet   S to N sample line ▢

Parish ▢   Field name ▢   Collector ▢

Date ▢   Start time ▢   Finish time ▢   Sieve mesh ▢

Soil:

Type

Clay [1]   Clay silt [2]   Clay sand [3]   Silt [4]   Sand silt [5]   Sand [6]

Colour

Dark Brown [1]   Brown [2]   Red [3]   Yellow [4]

Conditions:

Ground

Frosted [1]   Wet [2]   Sticky [3]   Moist [4]   Dry [5]

Cultivation

New harrow [1]   Weathered harrow [2]   New plough [3]

Weathered plough [4]   New shoots [5]   Post harvest [6]

Percentage of surface obscured/obstructed (Cover)

90-100 [1]   70-89 [2]   50-69 [3]   30-49 [4]   10-29 [5]   0-9 [6]

Slope

Flat [1]   South to North [2]   North to South [3]   West to East [4]   East to West [5]

Additional notes: ....
....

Stone   Enter number of stone. (Write *10+* if greater than ten)

Sample Line   For S to N: Line on the **W** side of box.   For W to E: Line on **S** side of box.

	Soil		Conditions					Stone			No. of bags	W to E sample line
	Type	Colour	Light	Ground	Cultiv.	Cover	Slope	Lias	Burnt	Other		
1												
2												
3												
4												
5												
6												
7												
8												
9												
10												

Figure 3.3. Fieldwalking record form, South Cadbury Environs Project

one collection strategy is being employed by the survey (intervals between lines; gridded collection; grab sampling etc.) that selected for each area ought to be specified. If photographs or drawings are made these should be listed, and a summary reference to the nature of any finds is likely to provide a useful cross-reference as bags of material accumulate.

Whereas fieldwalking was once regarded as a relatively straightforward matter of collecting or observing surface artefacts, then recording concentrations which were perceived as significant, most surveyors now view the process as an opportunity for the recording of much more information, some of which may provide qualitative or quantitative variables for later analysis. These variables may describe longer term elements of the natural landscape (topography, geology, soil type, aspect, proximity to water, etc.), so offering insight into the importance of location-types for past activities, or they may describe the particular conditions under which each unit is surveyed (ground cover, light, individual walker recovery rates etc.). In the latter case the data may be used to analyse distortions of artefact distribution and, because of their direct bearing on the rate of artefact recovery and observation, records should be made for every collection unit. Information about longer term characteristics will provide similar insights. Angle of slope will influence the degree of mobility of artefacts, as will soil type, which may also provide information about erosion. However, inferences about land use from surface soil data, or about the significance of water for the location of past activity should be treated with caution. Present soils have often been moved substantially, sometimes over a span of decades, and the hydrology of an area will also change.

The manner in which data are recorded is also very important. If fieldworkers are provided with generally headed boxes in which descriptions are to be written there may be an increase in the detail and, indeed, in the quality of conception of the particular place. However, there will be considerable scope for the vicissitudes of subjectivity, as well as greater difficulty in codifying the information for computer use (Figures 3.1 and 3.2). The alternative is to present a list of descriptive categories on the form which are ringed or otherwise marked (Figure 3.3). Here, too, there are problems. If the options describing visibility are given as "good", "normal" and "bad" (Carreté et al. 1995, 51) there remains considerable scope for variation of interpretation. It may be better to break down the options into more restrictive categories: Ground cover: 0-20%; 21-40%;.... Sun position: Overhead; behind walkers; to side of walkers; ahead of walkers; low;.... Cloud cover: Clear sky;... etc. Such options should reduce the effects of individual judgement, and further reductions are possible using diagrammatic representations of the optional parameters.

The minimum requirement for a recording form is that it should be capable of storing all data relevant to the questions addressed by the particular survey. Figures 3.1 and 3.2 are both examples of forms which require a considerable; amount of information, but differ in that the former allows the fieldworker to give more broadly interpretive accounts in boxes, whilst the latter gives a range of options from which selections may be made. To avoid limiting perspective there is a summary space which gives more latitude to personal interpretation.

3.3 Survey techniques and methods

There are a number of general survey techniques now available, some or all of which may be employed in a regional survey. The following list indicates the characteristic range of approaches: mapping; survey of the literature; aerial photography; topographical survey; recording of upstanding remains; geomorphology; environmental survey; geophysical prospection; surface/ploughzone archaeological sampling; intrusive archaeological sampling.

Every project ought to begin with a review of the existing literature for the target region, and with mapping. Thereafter the goals of the survey should determine the particular research techniques. Some surveys will have a very restricted function, such as the assessment of the condition of the physical archaeological record, perhaps accompanied by a conservation plan. An intermediate level of survey includes those designed to increase the number of recorded "sites" in a region. This range is likely to include analytical interpretation, and may go on to provide a further narrative, which ought properly to depend on the level of other information derived from detailed excavation records.

It is the principal goal of the most ambitious range of surveys to provide a narrative; the synopses and discussions of chapter 2 dealt almost exclusively with this group. However, the group requires sub-division according to individual surveys' chronological and geographical ranges. The chronological range of a survey such as that of the Biferno Valley (*4.4iic*) is vast; the geological and ecological description of the valley begins with its physical formation. For most surveys this would have formed a backdrop to human activity, but Barker includes this information as an integral part of the narrative. Whilst fully applauding this approach, I am employing the notion of archaeological survey narrative in a manner that restricts the range of my study to the post glacial period. It may well prove that future surveys will find adequate means for investigating pre-glacial human/hominid activity at the regional scale, but at present such work is either very much at the macro level (Newell & Constandse-Westermann 1996), or at the level of the investigation of a particular site.

The geographical range within a survey area pertains to the degree of inclusivity of diverse human activity for each determined phase. In the case of surface collection it refers to the frequency of sampling (for instance, of geological or ecological stratified sampling), and the degree of datable functional attribution which can be achieved curing analysis. Thus, a survey which uses a high sample frequency (say lines set 10m apart, a presumed 2m band of vision, and collection units at every 50m) will be considered to be in the upper part of the geographical range, so long as the material collected is analysed with due regard for its chronological specificity. If that level of analysis is not achieved the advantages of high frequency sampling will be lost.

At the beginning of this section I listed eleven technical categories which may contribute to a survey. A single programme of work within a region might include all categories, but I would suggest that any survey ought to include a minimum of three, two of which are necessarily mapping and a review of the literature. On rare occasions these last two alone may form a survey, but only where a full range of earlier fieldwork is available for re-analysis (for example, Larson et al. 1996).

The remainder of this chapter comprises an outline of each particular technical category, with examples, focusing especially on the use of six categories where resolution can be enhanced to great effect.

3.3i   Mapping

One of the first acts of regional survey must be to decide the extent of the target region. In many cases there will be existing records of finds or "monuments" which need to be entered onto a preliminary distribution map. Regressive mapping, the use of cartographic or written documents to reconstruct landholding and division over several centuries, is an increasingly common technique in British parish surveys (Mick Aston, and Jane Penoyre in Aston & Costen 1994, 19-44). It has been used to suggest that the outlines of estates established nearly two thousand years ago are sometimes susceptible to close analysis.

During the necessary desktop stage of a strata-based field survey, maps of the intended criteria may well determine the limits of the region, whether topographical, geological or ecological. Decisions about map scales ought to be determined by the nature of the research objectives, but all to often surveys have relied on small scale military maps (1:50,000), sometimes supported by aerial photographs. At scales smaller than 1:10,000 only the most generalised patterns of artefact distribution are visible; rectified photographic transcriptions are capable of representing generalised land division, but monuments such as barrows need to be represented schematically to appear at all. If such maps are used as a basis for fieldwalking the researchers are almost certain to use large collection units, usually determined by modern features such as fields, rather than regular grid references.

At a scale of 1:2500 it is possible to make detailed plans for line walking, and to present a reasonably full record of larger features noted during geophysical survey, and this order of scale is desirable for use in the field. However, representation of regional overviews require smaller scales if map sheets are to be of a practical size. A good example an effective compromise, using various colours and symbols for the sake of clarity, is the summary aerial interpretation of 450sq km around Danebury, Hampshire (Palmer 1984, internal pocket) at 1:25,000.

With the advent of Geographical Information Systems (GIS) for computers it has become possible not only to make a static model of a landscape in its various phases, but to carry out simulation experiments by varying climatic and other factors for each period. By virtual means mapping will soon become the prime interface between research archaeology and the general public.

3.3ii   Survey of the Literature

A survey of the literature may include primary texts in the form of documents contemporary with a research target period. It is sometimes a feature of projects dealing with estates, where maps and accounts may have survived. Secondary texts may include accounts of related research in monograph, journal or electronic form. Commonly, in England and Wales, it includes a trawl through the National Monuments Record (NMR), the Sites and Monuments Records (SMR), watching briefs and publications of earlier archaeological work in the region. When linked to mapping it may influence the size of region and mode and distribution of sampling; however, a recent assessment of evaluations in front of new development found that SMR entries "were not associated with significant archaeology", excepting those which included aerial photographs (Champion et al. 1995, 19).

Some surveys may be entirely desk based; a strikingly effective example was the review of archaeological work in the Great Basin, tested against hypotheses derived from detailed analysis of climatological records (Larson et al. 1996; see *2.3iii*, above).

3.3iii   Air Photography

Any effective aerial survey is likely, in practice, to include a substantial survey of the literature, including information derived from previous aerial photography. A combination or progressive erosion, changing patterns of settlement and woodland, underlying geology and soil moisture levels requires a corpus of prints over decades. Time of day (angle of sunlight) and the relative position of the camera to its target area on the ground are all critical factors. For continuing debates about technique and evaluation in Britain *AARGnews: The Newsletter of the Aerial Archaeology Research Group* (ISSN 0960-2852) is recommended; and for a general introduction, Rowan Whimster's slim volume for the RCHM(E), which includes examples from surveys (Whimster 1989).

It is probably fair to conclude that there have been two seminal air surveys in British archaeology, both covering central southern Britain: *Wessex from the Air* (Crawford & Keiller 1928) and Rog Palmer's survey of 450sq km of the Hampshire/Wiltshire border (Palmer 1984). The former showed archaeologists the range of evidence that the landscape might offer when viewed from the air, but it was Palmer's painstaking mapping and morphological analysis, coupled with use of existing excavational and geomorphological literature, which enabled one of the first truly archaeological aerial interpretations of a region.

Palmer's radical step was to arrive at a narrative by combining aerial data with a rich existing literature; a respectable, if more work-a-day *modus operandi* which continues to be used is represented by a survey of the Upper Thames Valley, which integrates existing documentation, such as RCHM and National Monuments Records with old and new aerial photographs in a single gazetteer for each mapped area (Leech 1977, 7-18). The gazetteer entries contain very spare chronological and functional evidence, derived from earlier fieldwork, but are analysed in terms of their archaeological potential. By the time of Leech's survey of the area several locations had succumbed already to gravel extraction (Leech 1977, 1), whilst others were threatened. The survey served to determine the intensity of archaeological work needed, and prioritised sites before their destruction. There was no attempt at synthesis, or the construction of a narrative for a "region" defined by geology which coincided with the goals of a modern industry.

Palmer made that step towards a regional narrative by employing much the same analytical process: a) *photo interpretation* - the subjective determination of what constituted a feature; b) *transcription* - rectified plotting of designated features onto a scaled map; c) *archaeological interpretation* - the use of extrinsic data, such as that derived from excavation, to phase features and even to offer suggestions about their functions (Palmer 1984, 5).

The survey was directed towards a particular chronological span, from Neolithic to Romano-British, with the Iron Age as the principal focus. After compiling a graphical and tabulated gazetteer of sites derived from aerial photograph- and documentary evidence (i.e. excavation records, etc.; Palmer 1984, 19-26), Palmer defined a set of morphological groups. Whilst these included the various forms of barrows, Roman buildings, roads etc., of critical importance to understanding the region in which Danebury was set was the analysis of linears and enclosures, including hillforts. Although the analysis of ditch-types, enclosure and approach/ entrance features assisted chronological diagnosis, the primary method was direct comparison with excavation records. By noting alignments and possible instances of superimposition a considerable degree of phase discrimination was achieved.

The subsequent chronologically ordered interpretative maps and narrative summaries were "intended to illustrate the development of early human settlement in the area, insofar as it is currently known and understood" (Palmer 1984, 123). The narratives, although slight in detail, are sufficient to generate hypotheses or models which might be tested by work carried out at ground level. Such work, in a spirit of maximum return for minimum investment, might entail trial trenching and remote sensing techniques. Alternatively, as Cunliffe has chosen to do, certain sites might be selected for excavation.

Aerial photography plays a part in the early stages of most surveys, and in areas where mapping does not exist or is at too small a .scale it has frequently been used to chart the area of survey. In conjunction with knowledge from other ground-based research it can be used as a preliminary means for evaluating areas of particular interest. In the Fens, prior to each season in the field, variations in the tone of the soil and cropmarks were transcribed onto maps, and later were used in analysis, not only linking anthropogenic marks directly to artefact distribution, but also for assessing the degree to which soil movements had affected the visibility of "sites" as artefact concentrations (Lane 1992, 8; Hall 1996, 192-198 etc.).

3.3iv   *Topographical survey*

It has long been recognised that undulations in the land surface often conceal coherent patterns, susceptible to archaeological interpretation. In the past the available technology for carrying out and recording fieldwork (clinometers, levels, pencil and paper, etc.) placed practical limits on the extent of topographic survey. But with the advent of Electronic Distance Measuring (EDM) systems, which locate and record relative distance, height and angles with much greater speed and accuracy, horizons are expanding. Whereas pioneers of monument survey such as Leslie Grinsell were restricted to recording the dimensions of individual landscape features (his well-known survey of barrows in southern Britain), modern regional archaeologists have the opportunity to sample much larger tracts of land with the potential for creating digital terrain models (DTM) with the potential for draping other classes of data over them.

An extraordinary example of the effective use of new technology comes from Ireland. There a survey of the literature and aerial photography had covered an area of approximately 100sq km, centred on the multiple monuments of Tara. Eighteen months of survey time, over three seasons, were devoted to a detailed topographical survey of the Hill of Tara itself, an area of approximately 0.5sq km. In the sure expectation of further work on the site four permanent control points were established and a network of temporary control points derived from these became the frame around which a network of surveys were built (Kieron Goucher in Newman 1997, 245-250). These comprised a mesh of 20m grids which were also used for geophysical and geochemical surveys. The data ("approximately 62,000 points were logged") were processed on a GIS programme, which has produced a variety of overall and detailed 3-dimensional surface models. The quality of the modelling, supported by data from the other techniques and with the advantage of knowledge derived from excavation and surface finds, has made it possible to discover new monuments and to create an overall phased sequence.

A detailed analysis of Alfred's Castle, Oxfordshire, was based on 8000 data points collected using a total station, which as vector data enabled the production of georeferenced layers for graphical presentation. Aside from giving the enclosure outline and the multiperiod features within it allowed the accurate plotting of trenches (Lock 2003, 43-48).

The value of generalised topographical survey will often depend upon the processes the landscape has undergone. Erosion can either reveal shapes in the surface, or rub them out. Ploughing will almost invariably be damaging, but the Tara survey has shown that by careful manipulation of the data even very weak features, including some disturbed by ridge and furrow, can be highlighted by low-angle computer "light" sources.

3.3v   *Upstanding architecture survey*

The locating and recording of structures takes a variety of forms in survey. Its most limited role occurs in environments where conditions such as heavy vegetation make searching the surface for artefacts impracticable, or when large scale surveys need a manageable target where more detailed work can be focused. The recording of "camp" platforms has been a routine part of survey in the North American South West, often as an element in the analysis of seasonal settlement patterns. Recording the form, dimensions and distribution of prehistoric stone structures, and association of them with finds enabled chronological and social distinctions to be made in a survey covering some 200sq km along the Yinhe River, Northeast China (Shelach 1998). A series of purportedly synchronic landscapes were mapped to provide an account of evolving social hierarchy.

Catherine Gerner Hansea stressed the rigorously limited approach adopted by the Africa Proconsularis survey in Tunisia (Dietz et al. 1995, 180-182), designed to provide data according with the specific aims of the project. She emphasised the importance of swift scale drawing, and of the complete lack of any attempt to trace parts of structures which extended under the ground surface. Sites were often first noted as areas avoided by the plough "with architectural remnants strewn among bushes and a few trees". More durable *opus caementicium* structures sometimes survived to a height of a metre, and elsewhere orthostats which had once reinforced mud-packed masonry stood proud of the surface, outlining collapsed and buried walls. In some cases there was visible evidence of robber trenching. At least one photograph was taken "from the best possible angle to get a general view of the area" at every site, and diagnostic architectural elements were also recorded in this way. The subsequent extensive catalogue demonstrates that the team acquired a lot of information about the distribution of building materials, but much less about the dimensions of their structures. Due to the lack of excavation of rural buildings there was no local database which might inform an interpretation of the results (Dietz et al. 1995, 349-377), and dating was by association with finds found in the vicinity of their debris. Nonetheless, there was enough to information to diagnose and phase bathhouses, cisterns, dwellings, fortifications, olive presses and some graves.

The UNESCO Libyan Valleys Survey covered a huge area (75,000sq km. Barker 1997, 26) of inhospitable pre-desert, and one of the first objectives of reconnaissance by vehicle-based teams was to locate architectural features. Other ground based teams walked 200m-wide transects along the edges of *wadis*, and periodically took right-angled turns into the surrounding *hamada* to investigate strips several kilometres long. Once building debris or relict walls had been located the team would collect material from the surface around it for dating purposes. From this preliminary work provisional maps were made and certain areas were selected for more detailed work (Barker 1997, 27-35). All types of "major wall construction and other recurring architectural features" including "burial cairns and hut footings" were sketched, and "special constructional features" were described to enable their classification (see *4.1iii* below). More "complex composite structures" were recorded photogrammetrically from a tethered kite by use of 5m scales and the plotting of known points on the ground. In some instances topographic survey was also applied. In all seasons except the first some remains were selected for excavation.

Principle targets of the survey were various walls which did not form buildings. These were classified according to five topographic location types, subdivided according to their position in relation to the landscape features of the immediate vicinity. A chapter is devoted to the analysis of their classification which forms the core to much of the subsequent landscape interpretation (Barker 1997, 193).

The amount of time invested in the recording of architectural features varies according to the survey landscape and the questions addressed by a particular project. Hansen expresses frustration at the extent to which the strictures upon the survey techniques limited the interpretative potential of the data. Shelach, on the other hand, probably overemphasises the actual synchronicity of his structure distributions. Much more convincing are the arguments of Chris Hunt and David Gilbertson which integrate particular features into a wider system within the Libyan desert, precisely by adopting descriptive categories which lend themselves to interpretation (although the authors' claim that the "classification was explicitly descriptive and independent of any supposed function" is at odds with the table on the facing page, listing classes such as "barrages", "dams", "sluices" etc., all good examples of class B and, in some cases, C interpretation. Barker 1997, 192-93; Table 7.1).

3.3vi Geomorphological survey

No survey which aspires to providing more than a regional gazetteer of sites can afford to omit an account of the soils and their formation. A useful initial research phase, depending on local conditions, can be a review of air photographs which very often show old stream beds, alluvial and colluvial deposits, as well as underlying geology (Aston & Costen 1994, figs. 8.9-8.12; Jameson et al. 1994, 173 and maps 1-8). In arid, eroding settings it is often possible to study the surface (Foley 1981, 57), but in a region where aggradation prevails, or where a variety of factors have influenced soil formation, the soil must be studied in section. This may be achieved by several means, often in combination.

3.3via Auger boring

Boring with augers is the most economical means for acquiring a soil profile. A 25mm screw auger, applied at frequent intervals in a line or in gridded clusters, is capable of providing a general guide to strata, although it tends to distort the measurable thickness of layers. With a 100mm bit a core can be obtained; more accurate, it is also much easier to read. A core of this size will better represent larger soil components, such as small stones, but if very stoney layers are encountered it is unlikely to drill through them (Shennan 1985, microfiche 5:A3). The larger auger, therefore, is likely to fail in the face of larger alluvial gravels or flinty chalk deposits, although at Miwa, Japan, depths of between 1.5 and 3m were achieved in 26 borings, set in a grid, through strata which included gravels (Barnes 1993, 39). In common with findings elsewhere it was noted that "a full understanding and interpretation of core sediments is seen above all to depend on excavated stratigraphy" and the opportunity for direct comparison was taken, from which "many points of agreement and discrepancy" were noted. One of the discrepancies, surprisingly, was the discovery of a feature fill in a core, which was missed during excavation (Barnes 1993, 40-41).

A programme of fieldwork at the coastal site of Brean Down, Somerset, combined excavation, test pitting and auger survey within approximately 2ha. Although this is a small area it included several types of environment, from inland agricultural land, to sandcliffs, to the intertidal zone. A "limited number of easily recognised horizons, often with a repeated pattern" were correlated with a sandcliff section (Keith Crabtree, Bell 1990, 90) in a survey of around 100 borings set in blocks or lines, usually at 5m intervals. Although the technique tended to miss thin contextual lenses, it proved highly effective in demonstrating the extent of general activity associated with broad-band phases.

In general auger bores are useful only when they support more detailed methods of soil profiling, such as machine trenching or test pitting. By running auger drills away from a trench sectioning valley bottom alluvium, onto the lower slopes, it may prove possible to study the process of colluviation by observing the soil components common to both alluvium and colluvium. In short, augering is at its most effective when some of the target soil components are already known and so recognisable.

3.3vib Exposure recording

Although non-intrusive this method offers the advantages of trial trenching, while costing comparatively little in labour, and even less in machinery! The method is simple: if natural or anthropogenic processes have revealed profiles these can be scraped clean, drawn and photographed. The method proved of enormous benefit in the Biferno Valley (Hunt in Barker 1995a, 59-82), where the deep incisions of the river, caused by tectonic movement raising the land, provided a succession of long profiles through much of the valley. This information was supplemented by records from road cuttings, quarrying etc. The technique was applied to similar good effect in the Southern Argolid (Jameson et al. 1994, 176-189) where the frequent, often deeply incised valleys demonstrated the character and volume of the soil lost from the hills, often by human agency.

The chief drawback of this technique is that it is likely to be biased. In the Biferno Valley survey a map showing the exposure locations, illustrates the survey's dependency on a central watercourse (Barker 1995, 70, fig. 30). In effect the survey has been conducted with the grain of the geomorphology, rather than cross-sectioning it. The bias was in part mitigated by evaluating the contribution of the valley sides and hills to the sediments, but the products of such evaluations ought to be regarded as hypotheses awaiting testing by reference to better preserved soils on the hills, should they have survived. Often such valley bottom deposits may be the only evidence for the existence of loess, long stripped from the slopes (Barker 1995, 71).

The technique is not restricted to valleys. It has been applied very effectively in Fenland environments, where the frequent drainage ditches provide many opportunities for the recording of profiles of the once waterlogged strata they are helping to destroy (Hayes & Lane 1992; Lane 1992; Hall 1996). David Crowther and Charles French have suggested that inferences may he drawn from abutting peat exposures which may inform "judgement sampling of deeply buried waterlogged contexts" (Haselgrove et al. 1985, 68).

3.3vic Test pitting

A test pit differs from an auger core in so far as its object is the section which remains after excavation rather than the soil removed from the pit. The distinction is crucial if this much more labour intensive technique is to be justified. The four faces of a pit should provide an undistorted profile which is likely to include far greater detail than a core, allowing a more sophisticated model of soil composition. It has the further practical advantage that only in the most extreme cases will layers prove to be impenetrable.

Because the target is the section, the pit can be dug without being impeded by the close observation of the spoil, unless it is in an archaeological sensitive area; equally, a mini-mechanical digger may be used.

The effectiveness of the technique depends on the quality of the recording (of special importance in understanding the movement of soil is particle size, which conveys information about wind and water caused erosion) and frequency of sampling. Outstanding results were achieved at Brean Down where the technique was combined with augering (see *3.3via*) and excavation. The layout of twelve pits was set within the framework of the auger survey (Keith Crabtree in Bell 1990, 94). Three pits were machine-excavated and samples, including artefacts, were taken from their sections. The other pits were dug by hand, their finds recorded three-dimensionally. Most pits reached a depth of 2-3m.

It is only in the past three decades that erosion and the laying down of colluvium have been appreciated as widespread processes (Bintliff 1992), particularly through the extensive survey work in the Mediterranean. Economical techniques for evaluating their impact at a regional scale remain problematical; extensive test pitting is extremely labour intensive. Yet it is only by acquiring information about localised geomorphology, within areas where surface collection/observation or ploughzone sampling are to take place that we can make a realistic assessment of the significance of the recorded artefacts.

The South Cadbury Environs Project has preferred a more extensive approach, digging one 1m square test pit to natural at every hectare as part of an integrated plan with geophysical survey and shovel pitting. It has proved an effective means for judging the efficacy of the complementary techniques whilst identify traces of probable settlement activity, particularly from the prehistoric periods, where there has been no evidence of it on the surface. It has proved particularly effective in identifying general areas of activity from later prehistory, a period when pottery, the principal tool for relative dating, is very friable and has a poor survival rate on the surface. In any event finds of the period are often almost inaccessible, concealed beneath a metre or more of alluvium.

3.3vid Machine trenching

The most prominent advocate and practitioner of this technique in Britain is Martin Bell, who broke new ground with his work in Hampshire and Sussex valleys. At its crudest a long trench (20-30m) is machine excavated to the base of alluvial or colluvial deposits, and then fully recorded. Bell added to this simple approach the excavation by hand of one long section. Thus the diggers worked with a clear view of the stratigraphic position and were able to add archaeological and environmental data to the geomorphology.

Until this method was introduced the number of artefacts recovered in augering and test pitting was so limited as to make reliable association of sediments with archaeological periods very haphazard. Work at Bishopstone (Bell 1977) and Chalton (Bell 1983) was not only able to link episodes of erosion to phases of prehistoric agriculture, but was able to demonstrate the importance of singular events, as well as long-term climatic conditions, in the erosion of cultivated landscapes by analogy with modern observations (Bell & Boardman 1992, 21-33).

While earlier studies of his type can usefully inform analysis of the results of test pitting, only further machine trenching can offer a data-set to provide this quality of information within any specified region. Indeed, as has been noted elsewhere (Bell 1983), there can be sharp variations in results from one valley to the next, even where the geology and topography appears similar. However, trenching probably represents the most cost-effective technique from which regional generalisations can be made, since the large sections will show sediments which have accrued from an extensive portion of the landscape.

3.3vii   Environmental sampling

A representation of the natural or anthropogenic environment during targeted periods in a study region is highly desirable, but has often proved elusive. Although paleobotany and palynology have become increasingly sophisticated, the quality of the analytical procedure is undermined by the sporadic availability of the evidence. There are, of course, notable exceptions in areas of wetland, but very often projects have been forced to rely on very localised deposits from which far reaching claims are made for much larger and varied landscapes.

Other, more widespread, sources of environmental data are watercourses. Even in semi-arid landscapes botanical remains may be sealed in alluvial sediments; in practice, however, there are often difficulties in dating them. Carbonised remains will occur in a wider range of contexts, but represent a different class of material, most commonly restricted to domestic crops and associated weeds.

Faunal environmental data are even less suited to extrapolation across a region, although this has not prevented authors from attempting it (Barker 1995, 149). Domestic faunal and carbonised remains, by virtue of being the direct result of human subsistence or economic activity, if undisturbed by later taphonomic processes, usually occur in association with other datable cultural remains.

Excepting those areas of wetland where widespread preservation of pollens is the rule and the collection of cores becomes practical, recovery of environmental data tends to be an incidental by-product of larger scale intrusive geomorphological or archaeological techniques such as machine trenching or full scale excavation. In the Sergemes Valley (Dietz et al. 1995, 90-91), in addition to excavation and coring, samples were taken from exposures in the banks of *oueds*, sharply defined river valleys prone to rapid erosion when their small streams are transformed into torrents by heavy rainfall and the runoff from the hills above the upper valley. The complex geomorphology generated by repeated episodes of violent flooding and aeolian sedimentation made stratigraphic dating of the material difficult (Dietz et al. 1995, 89), and the amount of organic remains in the sediments was usually too small for radiocarbon dating.

A different method of sampling was employed in a comparable North African environments by the UNESCO Libyan Valleys Survey. Trial trenching was specifically targeted in narrow slots across middens. The arid conditions prohibited the use of a flotation system, and because removal of soils was not practical dry sieving was carried out close to the point of recovery. Usually the soil was dry-sieved through a 1mm mesh, then a sample of the soil that had passed through was re-sieved through a 0.5mm mesh. Only in the Final year of fieldwork (1989) was an archaeobotanist able to be present to supervise the process, so that only the data for that year could be relied upon to provide a measured sample for valid statistical inferencing (van der Veen, Grant & Barker in Barker 1997, 230). Nonetheless, all the recovered material contributed to valuable insights about hitherto unknown Romano-Libyan crop regimes and animal husbandry.

In an attempt to place later prehistoric and Roman agriculture in the northeast of England within a regional perspective Marijke van der Veen arranged for the collection of carbonised remains from all excavations, including rescue, over a six year period (1981-87). Where excavated areas were small enough directors were expected to collect samples from every context. On larger excavations probabilistic sampling strategies were used, on one occasion supplemented by judgement samples. At eight sites 30 litres soil samples (two full buckets of a standard size) were retained from most "well-defined, sealed" contexts (van der Veen 1992, 21), from which carbonised matter was extracted by flotation. Nine sites were listed with general descriptions of the site overall, including basic plans and interpretation of some features and dating (van der Veen 1992, 29-50). The environmental sampling strategy employed at each site and the environmental data retrieved from flotation are described.

The quality of the subsequent interpretation demonstrates the importance of the involvement of the archaeobotanist's direct involvement at least with the excavation directors - a point emphasised by another archaeological scientist, Tjeerd van Andel, who has called for an interdisciplinary as opposed to multidisciplinary approach (van Andel 1994, 28). He asserts that a full engagement of the different disciplines with each other at every stage in the project will maximise the return from the various techniques employed.

At present the processing of some classes of environmental data is considered to be expensive, but it provides an invaluable insight into past modes of subsistence and economy (see *3.3viiia* and *4.1iv*). Regional archaeology generally has still got much to learn about how best to sample large tracts of land to benefit from this sort of material.

3.3viii   Geophysical prospection

Non-intrusive techniques for representing subsoil features have shifted from being esoteric experiments (Clarke 1990, 11-26) in the 1950s and 1960s, through a phase of uneasy acceptance and wider application in the 1970s, to a generally established element of archaeology in the 1980s. Early sorties in the field were hampered by manual data recording followed by a choice of manual plotting or the use of crude, memory-limited computers at some distance from the site. Consequently, geophysical prospection was carried out on small areas, usually comprising meshes of no more than a dozen 20 x 20m grids, focused on known sites, resulting in its under-representation in regional survey literature. Where it does occur, it is as a secondary tool investigating narrowly targeted areas on the basis of fieldwalking or aerial photography, usually as a means for honing decisions concerning the location of trenches.

Work carried out with a range of instruments including the Soil Conductivity Meter and a Proton Magnetometer at Cadbury Castle from 1967-70 broke new ground by covering the entire interior of a hillfort (some 7ha), and generating some very good quality data (Alcock 1972, 54-62; figs. 6 & 8). Hand-plotted dots by Chris Musson, offered a break through in presentation. Particularly striking at the time was the high level of correspondence between the electrical and the magnetic survey (Alcock 1972, 62). This success seems to have been a false dawn; ten years later meshes of fewer than a dozen 20 x 20m grids were being employed over a very limited number of selected sites in the Biferno Valley. The graphical presentation was crude and far from user-friendly (Barker 1995a, 93-9; figs. 56-59).

With the advent of data loggers linked or built into the instruments, and the ability to transfer data directly to microprocessors, the scope of geophysical prospection has been transformed, although many archaeologists have been slow to realise the increased potential.

The bias towards "site" which governed the choice of seven selected areas for intensive fieldwork, chiefly excavation, for the Danebury Environs Project (a very brief outline appears in Cunliffe 2000, 14-16), led to the use of geophysical techniques as a prelude to excavation. Thus, for instance, a complete banjo enclosure at Nettlebank Copse was surveyed, but the area around it, its context in the wider locality, was excluded. The only exception to this practice was the large scale survey targeting field systems at Windy Dido, which has enhanced the view of a probably Bronze Age agricultural landscape. Barry Cunliffe has also been the instigator of the Wessex Hillforts Project, a programme surveying the whole of hillfort interiors on chalk. The relatively compressed span of prehistoric activity within such monuments allows the possibility of strong hypotheses concerning their internal structures.

The RCHM(E) has commissioned surveys in various parts of the country, but once again they have focused on known monuments, rather than landscapes (much of this work, carried out by Geophysical Surveys of Bradford, has been extremely effective, but remains unpublished). What remains to be addressed is a satisfactory means tor sampling landscapes. Experimental methods have included the use of narrow transects (20 to 40m wide) and small, randomly distributed grid meshes. Neither have proved very effective. On suitable geology there has been no difficulty in discovering distinct archaeological anomalies; the problem is that the sampled area is too small to reveal meaningful, morphologically analysable patterns except where trial trenching can provide dating and functional evidence.

These approaches are the consequence of a perceptual error; in a regional context, the discovery of "sites" should be regarded as a desirable, but secondary by-product. The issue is one of scale, not from the point of view of limits on the resources for extensive surveying, but in terms of the represented object. The computer generated plot of a survey over a wider area may best be compared to an aerial photograph, but often with much greater resolution. Contrasting patterns of alignment and morphological character can be distinguished and mapped as separate systems of anthropogenic landscape structure (Tabor & Johnson 2000, figs. 3-9). In addition to large scale landscape features such as ditches, strong isolated anomalies may represent kilns, or hearths within structures which otherwise have insufficient magnetic strength to register.

Although the present state of technology, and the limited financial and labour resources available, make the surveying of whole regions improbable in the near future, new dual system gradiometers with much bigger memories (i.e., the Geoscan 256 or Bartington 601-2) make the surveying of areas approaching four hectares in a single day perfectly feasible. By focusing on localities with meshes of several hectares at a time, geophysical prospection will assume a more appropriate role in regional survey. It is instructive to compare the time taken to cover similar areas with fieldwalking.

An increasing number of techniques are being tested, but only a few have reached the stage where they are suited for inclusion in programmes of regional survey. Of those I have chosen to exclude, Ground Penetrating Radar (GPR) is the technique most likely to be appropriated in the future; already it has been used as a targeted technique for increasing the resolution of specific areas within larger scale surveys, notably for enhancing images of structures in ancient urban environments, and for creating pseudo-section profiles; its most important application is in the area of deep stratigraphy. But at present it remains a slow and comparatively expensive procedure, a point illustrated by the fact that an area of less than 3ha at Wroxeter represents the largest continuous area ever surveyed (White forth.).

Resistivity has been considerably slower, but a new instrument from France, used at Wroxeter, has shown that it may well become as quick as gradiometry. The new technology is not generally available, but older instruments already offer a reasonable rate of progress, if tightly targeted.

Recent work at Tara, Eire, has demonstrated how the integration of all these techniques, mapped onto 3-dimensional computer representations of a detailed topographic survey can produce a stunning amount of information about the structure of a landscape. Since this information is distributed throughout the report (Newman, 1997) no specific reference is given here. But every archaeologist wanting to assess the quality of information which can be extracted from this approach to a locality or region should read the book!

3.3viiia Magnetic susceptibility and phosphate analysis

A form of environmental sampling which is becoming more popular in regional projects is geochemical survey. This is frequently placed in reports under the heading of geophysical survey, but while they may locate features they are better indicators of general patterns of the distribution of certain categories of environmental data. The most widely used targets for analysis at present are the phosphorous content and magnetic susceptibility of the soil. Both techniques are applied because of their potential for picking up traces of organic remains. In effect they will give some clues to the intensity, and sometimes, the nature of activity associated with human agency, especially high concentrations of excreta, which may result from the enclosure of animals. Magnetic susceptibility will also show thermo-remnant magnetism. Samples can be obtained relatively easily with a hand-held auger, within formal grids or lines (for good descriptions of the procedure see Newman 1997, 287-290), or from excavation. At present sampling at every 20m in a line is considered acceptable to discover general patterns, with the frequency often being increased to every 5m where more intensive activity is known, and sharper variations in readings may be expected.

Magnetic susceptibility can be measured in the field with a coil, or it can be integrated into a test pitting/ augering programme. At South Cadbury Duncan Black explored the relationship between geophysical anomalies and excavated

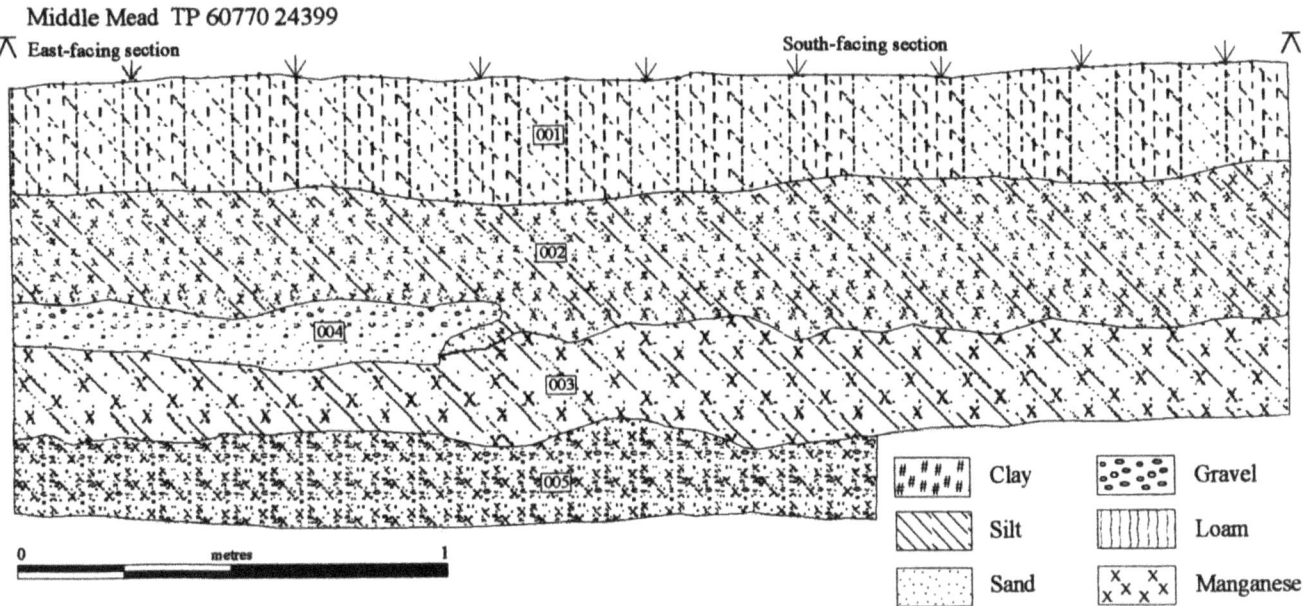

Magnetic susceptibility readings by context (two samples taken from Context 002, one above other).
Context 001: 21.5k; Context 002a: 29.7k; Context 002b: 41.1k; Context 003: 20.0k; Context 004: 677.6k; Context 005: 42.1k.

Figure 3.4. Magnetic susceptibility sample readings by context showing a a probable anthropogenic context (004) which was not recognised as such at the time of the test pit's excavation.

contexts in targeted test pits. He took surface readings at every 10m, and collected ploughsoil samples at every 20m on the same grid. These were compared with the magnetometry plots, and then with samples from each context in targeted test pits and those dug at regular intervals (1 per hectare). The most useful results came from these stratified samples. In some instances test pits revealed no contexts thought likely to correspond with geophysical anomalies at the time of excavation, which nevertheless proved to have a diagnosed contexts which proved to have very distinct magnetic signatures. Equally, where deep contexts had more than one sample collected magnetic susceptibility provided evidence for a changing context composition over time (Figure 3.4. Note contexts 004 and 002 in particular).

A much more intensive, intra-site programme of magnetic susceptibility testing, with sampling at every 1m during excavation in a grid laid out over the remains of two circular structures at Brean Down, Somerset (Bell 1990, 197-202), provided evidence for the distribution of activity in time as well as space, so offering a much fuller than usual account of the use of a building (for a cautionary note see Evershed et al. 1997 and *4.1iv* below).

3.3viiib *Magnetic survey*

At present survey by gradiometer represents the fastest surface-based form of geophysical prospection, providing the underlying geology is suitable. The instrument has, over the past two decades, virtually replaced earlier magnetic meters such as the proton magnetometer. By incorporating two fixed spaced sensors in the vertical plane the gradiometer has muted the effect of diurnal and regionally specific influences, giving far more accurate data sets relating to subsoil features. This preference contrasts with the late 1970s when, in the Biferno Valley, both proton and differential magnetometers were found to be more effective on "complicated sites with weaker variations" (Barker 1995, 53). However, the former have been used recently at Wroxeter, over small areas, where it is also hoped to use a "caesium magnetometer since this is a more sensitive instrument in detecting traces of timber buildings" (White forth.). Hand-held, the gradiometer records fluctuations in the polarity and intensity of the soil's magnetic properties. These are altered in particular by intensely heated objects (giving rise to thermo-remnant magnetism) and by the decay of organic matter. Typically magnetic survey is good at identifying ditches, hearths, kilns, large pits and midden deposits. It is often claimed to be less effective for identifying stone structures but survey's such those at Wroxeter and Bradford-on-Avon (Figure 3.5) show that in favourable geological conditions it can still be highly effective. Magnetometers are limited in their effectiveness where underlying geology comprises igneous material with variable ferrous magnetic properties, and responses tend to be weak on clayey soils.

Until recent technological developments increased memory capacity the frequency of sampling had to be played off against the time taken. Thus in the mid 1990s South Cadbury Environs Project decision to survey a field of 18 ha with readings taken at every 1m along traverses set at 1m intervals within 20m grids was a substantial undertaking for the then amateur study, relying entirely on volunteer's free time (Figure 3.6). At this level of resolution boundary ditches and possible walls, barrow ring ditches, burnt deposits and, with variable distinctiveness, Roman building walls were identified, and even tentatively phased. Since two readings have been taken per metre, and since 2002 at 4 or sometimes 8 per metre.

The depth range of standard gradiometers is stated to be around 1m, which would often leaves features sealed by alluvium obscure. However, recent work by at South Cadbury has demonstrated that black soils forming Romano-British upper ditch fills were detectable under 1.5m of colluvium

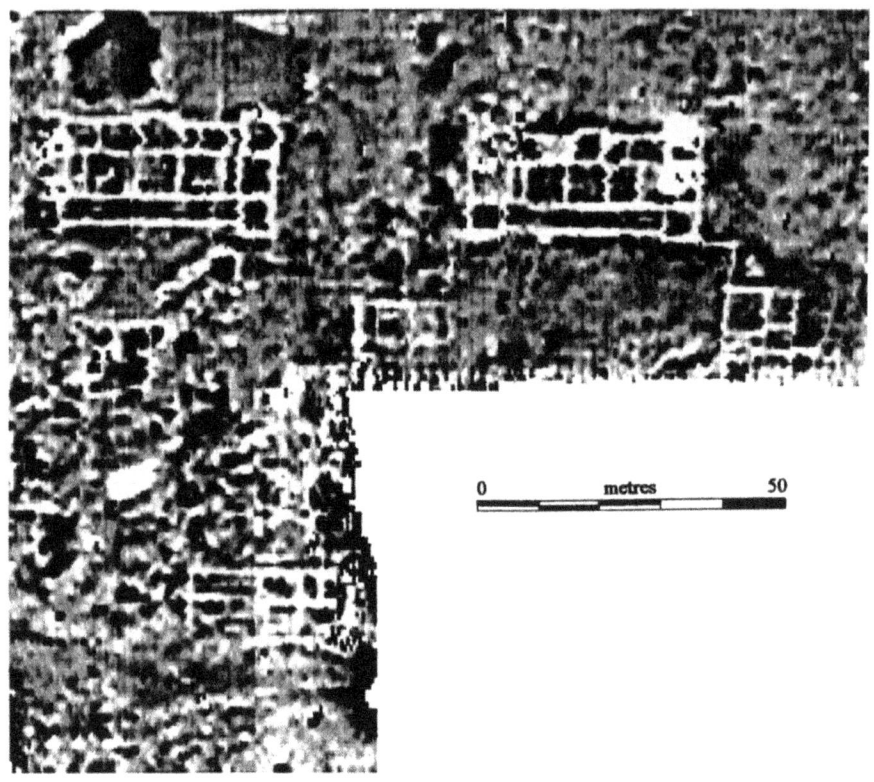

Figure 3.5. Roman walls detected by a gradiometer (Geoscan Fm36), Bradford-on-Avon, Wiltshire

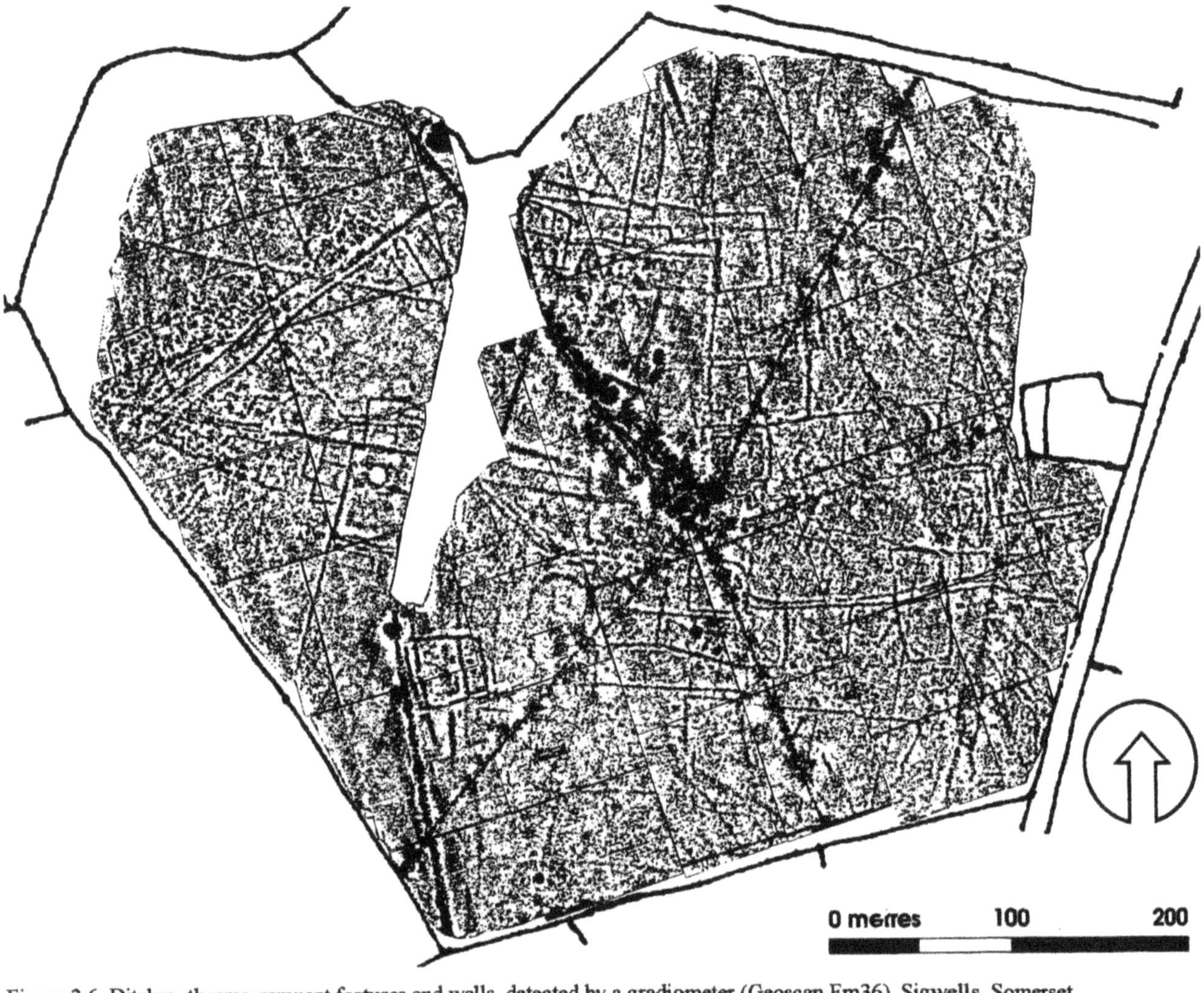

Figure 3.6. Ditches, thermo-remnant features and walls detected by a gradiometer (Geoscan Fm36), Sigwells, Somerset

The Geoscan FM 36   The Bartington Grad601-1

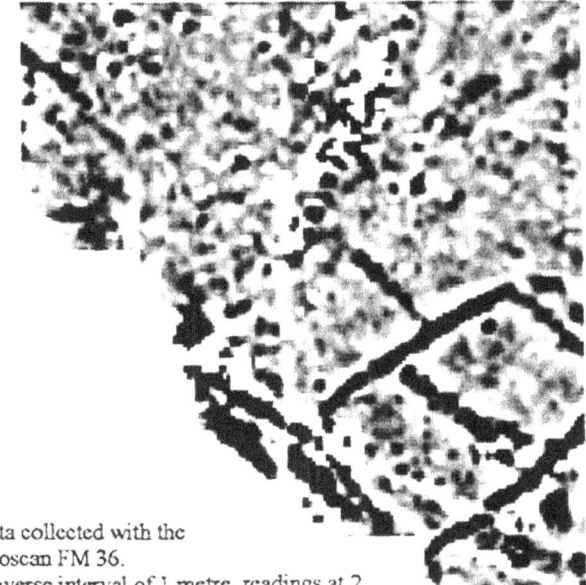
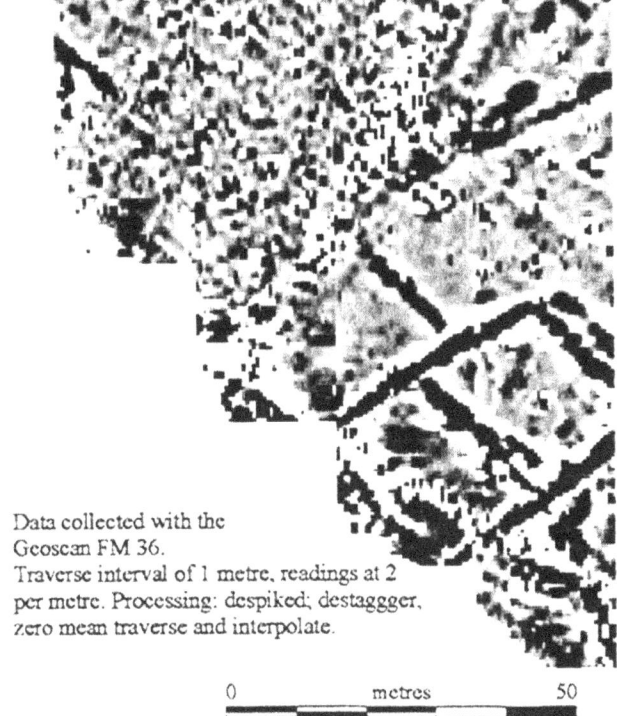

Data collected with the
Geoscan FM 36.
Traverse interval of 1 metre, readings at 2
per metre. Processing: despiked; destaggger,
zero mean traverse, interpolate and low pass.
(A vertical white stripe across the centre of the plot
is due to collection error independent of the instrument).

Data collected with the
Geoscan FM 36.
Traverse interval of 1 metre, readings at 2
per metre. Processing: despiked; destaggger,
zero mean traverse and interpolate.

0   metres   50

Figure 3.7. Comparison of depth ranges of gradiometers, Crissells Green, South Cadbury.

using a Geoscan FM36 (Tabor 2002, figs. 5.39 and 5.41), whilst the Bartington 601 appears to achieve responses at an even greater depth (Figure 3.7). However, it is severely hampered in areas where there are extensive underground services such as cables and pipelines, which include ferrous magnetic components.

At Wroxeter detail has been enhanced by the application of 0.25m sampling intervals with a Fluxgate Gradiometer (White forth.), but the apotheosis of site-based gradiometry has been achieved at West Heslerton, where one of the largest programmes of excavation in Northern Europe was carried out from 1977 to 1995 (Lyall & Powlesland 1996, 2). The excavation covered 20ha; a complete gradiometer survey of the site (unfortunately the data from one area of about a third of the total proved unusable) was used to forewarn the excavators of the location of features. On chalk, the technique proved highly effective, but the complexity of multiphase features presented problems of sequence interpretation. It was found that by removing the topsoil and increasing the sampling frequency to 4 per metre, more, better defined features were discernible. On the basis of their relative magnetic values it was possible to estimate the relative phasing of different features, and the authors claim that this analysis informed the stratigraphic interpretation of features following excavation (Lyall & Powlesland 1996, 4).

3.3viiic   Electrical survey

The measurement of electrical resistance in the soil is, in effect, a means of gauging relative moisture levels. Features such as walls and stoney banks offer high resistance due to low moisture levels, whilst cut features, although there are seasonal variations. Paul Young, University of Keele, has been reported as suggesting that, at Wroxeter, "continuous monitoring of the resistance on the site should be established for one year to investigate any electrical variations that may occur over a period." Such work might have more general implications concerning the optimum times of year for resistivity survey (White forth.), although variable climates make this a far from routine matter.

Data values may be heavily influenced by geological variation, and by prolonged dry spells or heavy rain during a survey may also present problems. In the Biferno Valley survey, carried out during high Summer, magnetometry was preferred because of the extremely dry conditions (Barker 1995, 51-53) but, by contrast, in the Tarragona survey, where magnetic responses were considered weak, resistivity proved effective in locating the shape and extent of buildings (Carreté et al. 1995, 224).

The application of resistivity meters as regional tools has continued to be limited because data collection is slowed by the need to insert mobile probes into the ground at every sampling. This is a much slower process than gradiometry because of the physical contact with the soil. However at Wroxeter, in the project's pursuit of experimentation, a method using "a multi-probe, automated, continuous-reading resistivity meter" has proved capable of providing a much faster survey, with excellent results (White forth.). The present system, comprising low-based rotating probes, is only suited to flat ground with low vegetation cover.

The increased speed of the new method opens up the possibility of use much in the fashion I have proposed for gradiometry. In areas high in ferrous magnetic disturbance widespread use of resistivity meshes would very often be of greater value. However, for surveys of extensive rural landscapes the nature of the target population must be considered. If cut or heat-derived features are anticipated magnetic survey is likely to remain the preferred technique for large scale use, with resistivity targeted subsequently, where morphological analysis suggests the likelihood of stone structures. Only where a deserted urban site is the object, is an electrical survey likely to be the most appropriate technique for widespread use, although the proximity of electrified rail systems can rule it out altogether.

The use of the technique for individual or clustered stone founded/built structures where concentrations of finds occur will no doubt remain a prevailing practice in the regional scheme but, where feasible, it is highly desirable that smaller resistivity grid meshes should be mapped onto a background provided by a magnetic survey mesh. On the one hand the anomalies revealed by one technique may be thrown into sharper relief by the other (Figure 3.8), but it is important to be aware that something shown by a gradiometer may not show in resistivity, and vice versa.

Integrated in this way, a harmonious representation of structures and boundaries is more likely to emerge, displaying patterns of synchronic and diachronic development of settlement and land division which can be tested against ploughzone finds and small scale invasive techniques.

3.3ix   Non-intrusive collection techniques

Techniques for collection which do not disturb archaeological deposits other than those which are routinely or drastically disturbed by recent or contemporary activity are described as non-intrusive for the purposes of this section. They also include the various methods of surface collection - the systematic counting, recording or collection of surface material while walking in a mapped area (fieldwalking/ pedestrian survey). Within this group I distinguish between two general categories of surface excavation: 1) Ploughzone sampling - must meet two criteria: a) that the target area is or has been ploughed, and that there remains a distinctive ploughsoil horizon, even if it has since been sealed by a turf; b) that only the ploughsoil horizon is sampled. 2) Topsoil sampling - a technique used either where the history of the subturf soil formation is poorly understood, or where ploughing is thought unlikely to have occurred, but where ancient surfaces have been concealed by natural processes of aggradation by means other than the movement of soil (i.e., the rotting down of organic matter in scrub or woodlands etc.). In terms of their application in the field the two are often identical, and as a technique may be described as shovel pitting, but due regard ought to be given to the different character of the erosion which will have acted upon the two soil mediums. It has been asserted that a regularly ploughed soil becomes "homogenous" (Steinberg, 1996), implying a generalisation of the discard pattern. Although topography will have some impact upon artefact movement, some research into the subject suggests that only major erosive incidents will have a significant impact on distribution. Where erosion has not been a major factor, and the soil has been unploughed, the relationship between the place of discard and the collection unit may be much closer.

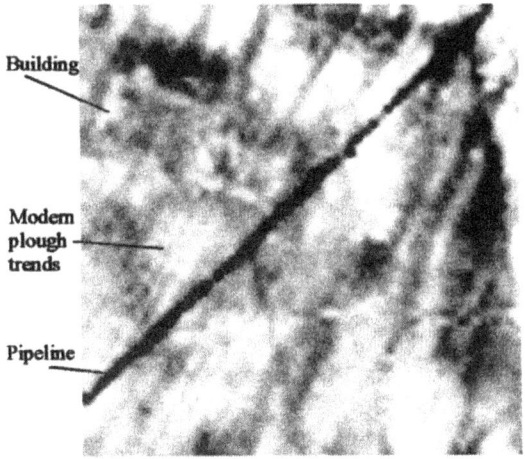

Resistivity plot (Geoscan RM 15)

Gradiometer plot (Geoscan FM 36)

Figure 3.8. Comparison of resistivity and gradiometer surveys, Sigwells, Somerset. The modern iron pipeline is clearly visible in both plots, but it has much less impact on the resistivity, so that an annexe to the Romano-British building is clearly visible. It should also be noted that several other features indistinguishable within the complex gradiometer plot are well defined by resistivity.

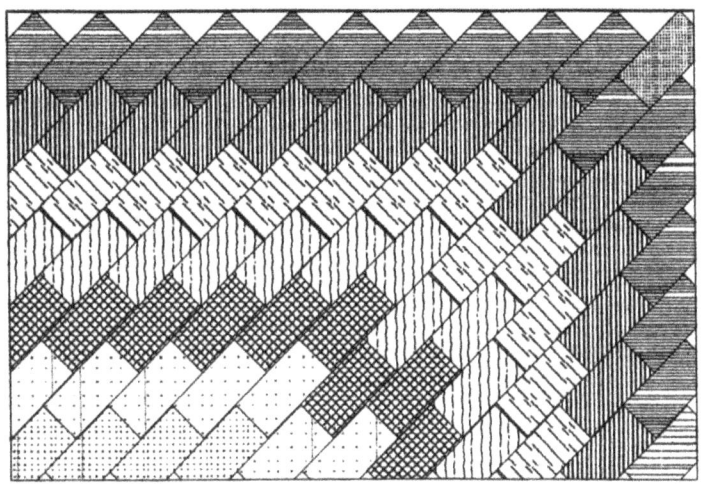

Figure 3.9. Segmented study area grid, Amboseli Park, Kenya

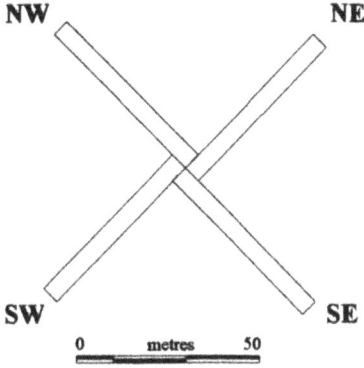

Figure 3.10. Sampling unit, Amboseli Park, Kenya

47

In ploughed fields the target population of finds is broadly speaking the same for surface collection and ploughzone sampling, since the artefacts are held in the same medium, the ploughsoil. Of course, much fieldwalking takes place in arid and semi-arid zones, some of which may not be cultivated; but these remain comparable where they present surfaces routinely changed through various forms of erosion, rather than aggradation, so that they offer a similar class of evidence. Nonetheless, there is an important distinction to be made: a cultivated soil exists as a 3-dimensional entity, whereas an eroding surface must be treated as 2-dimensional, since it represents an exposed, already truncated, plane of sedimentation. As a consequence, ploughzone sampling, a procedure by which the soil is sampled in both vertical and horizontal dimensions, may be inappropriate in an eroding environment since penetration of the surface might interfere with a stratigraphic sequence.

For theoretical purposes a weathered, cultivated surface and a naturally eroded surface offer a class of data comparable with any other information retrieved within the vertical span of the ploughblade, which over several years produces a homogenous soil. However, it should be noted that in a landscape eroded by wind it is very probable that a higher percentage of the artefacts, discarded over long periods, will be present on the surface. In a field which has been ploughed repeatedly there will be a statistical correlation between the number of artefacts on the surface, and the number concealed in the ploughsoil, although this will vary according to soil conditions, topography, climate and artefact characteristics. In theory this correlation should be revealed if a sufficiently large surface area can be sampled by fieldwalking, and then excavated and sieved in units based on a volume of soil large enough to produce a sample of artefacts of sufficient size to be statistically viable. However, in most cases the volume of soil to be sieved would be too great to effect a direct correlation economically.

Fieldwalking has been a preferred technique of regional surveys for two reasons, the foremost of which is its economy, and the second of which is its non-invasive, and presumed repeatable, nature. Ploughzone sampling and shovel pitting flout the first of these of these principles, requiring the effort of removing soil. Whichever terms are used to describe these latter two activities, they are liable to be misunderstood. A procedure labelled Test Pit Sampling (TPS), defined as "examination of a set of subsurface tests, most often located within larger, well-defined sampling units, in efforts to discover previously unknown archaeological deposits" has also been called "shovel probing" and "shovel pitting" (Nance & Ball 1986, 458). Praxis demonstrates that each of these terms has meant different things to different researchers so I have chosen to take as a definitive guide my linguistic sensibility as a Somerset speaker of the English language! To me a "test pit" must be a fairly substantial (girt) hole which cuts through more than one stratum (i.e., more than merely the plough or topsoil), so it seems inappropriate to use that term to cover non-invasive techniques. "Shovel testing", or "shovel pitting", to my ear, implies superficial or shallow digging, and is a usefully non-specific term to cover the testing of top or ploughsoil.

The rationales for the use of each non-invasive technique are set out in the following subsections. Whichever is used, it is essential that every survey ought to have a clear policy concerning exactly what classes and types of artefact it expects and intends to retrieve, whether it be "total" or selective collection. Although the policy will affect field strategy, the choice of the range of artefacts for collection should be based on the narrative aims of the project and will be discussed below (see *4.1*). All forms of ploughzone or surface sampling ought only to be used in conjunction with geomorphological or archaeological test pitting to understand how the structure and depth of soils will effect the perceived significance of distributional data.

3.3ixa Fieldwalking: generalised regional

Fieldwalking developed as a means for locating sites at a time when that term was treated as all but synonymous with "settlements". Often it was carried out by amateurs who submitted their finds to local museums along with map references of varying degrees of approximation. In Britain a substantial number of entries in Sites and Monuments Records and National Monuments Records are based on this information. Eventually it became a means for fairly systematic research employed to produce site distribution maps or as a means for assessing the archaeological potential of areas threatened with development. Nowadays it may still be used as a mode of reconnaissance in very large scale surveys, such as those in the North African desert.

Most fieldworkers recognised the biases inherent in what were usually subjective evaluations concerning whether or not a scatter of finds implied the existence of underlying structural remains, but there was a pessimistic assumption that this was the best that could be achieved (Hope Simpson 1983, 47). Fortunately for the discipline it was noted that if the intervals between individuals in a team of fieldwalkers was reduced the number of diagnosed sites increased; significantly, though, it became possible to discriminate between sites on the basis of the extent of the surface scatter. This simple fact sharpened debate about what volume or proportion of artefacts constituted a site and, inevitably, what was the significance of a scatter of artefacts, or background noise, which was not considered to constitute a site.

While some archaeologists had accepted, since the 1930s, that low frequency scatters might represent manuring zones, there was a growing recognition by the 1970s that variations in density of occurrence might offer deeper insights into the wider distribution of human behaviour.

One of the earliest thorough attempts to exploit this concept was Robert Foley's survey of the Amboseli National Park (see Chapter 2). The survey crystallised several evolving concepts of discard behaviour:

1) that environmental factors contemporary with the discarding of artefacts will influence their distribution;

2) that the artefactual record is continuous;

3) that behavioural differences will have caused variations in discard intensity.

Foley went on to consider how to adapt his field and analytical methods to maximise their usefulness and validity. To achieve this he required contextual information and one of the reasons that he chose to work in the Amboseli was precisely that the information he wanted already existed. The year before he started work a survey of habitats in the

Park had been published (Western 1973). This provided estimates of the number of animals per species which could be supported by the present conditions of vegetation and hydrology in the area. By using existing environmental data it was possible to reconstruct earlier environments, and so to estimate the biomass available to human populations within each chronologically and geographically defined ecological unit (Foley 1981, 39). Using ethnographic models applied to the available food resources, Foley made predictions about patterns of discard behaviour and from these was able to suggest expected proportions of artefact density from one ecological unit to another.

Before the survey proper began other factors had to be considered, one of which was the present condition of the land surface, since this might influence the perceived distribution pattern. Thus, prior to full scale survey, Foley carried out experiments in collection strategies. The continuing erosion of most of the survey area's surface suggested that conditions for collection would generally be good, providing that vegetation or compaction did not obscure artefacts. To test this Foley set out several $4m^2$ grids. After picking up surface artefacts, he sieved the soil in each grid to compare the rate of surface recovery with that for the topmost layer of compacted soil. In all but one grid the number of surface finds exceeded those retrieved by sieving, indicating that compaction had little impact on artefact visibility.

The size of the area, some 600sq km, precluded any ambition to survey the whole. However, Foley wanted as wide and as representative a sample as possible. To this end he divided the area into 4 x 2km segments which were aligned roughly perpendicularly to the prevailing ecological grain (experiments with ecological surveys elsewhere suggest that a cross-section yields the most representative species sample). Within each whole segment four sampling points were selected randomly (Foley 1981, 34-35), with two further points held in reserve in case any of the four selected were unsuitable. Each sampling point comprised a cross of four 50m x 5m spokes, thus yielding a coverage of 4000sqm or 0.05% per segment (Figure 3.9 shows the segmented mesh, and Figure 3.10 shows the plan used for each collection unit).

By this method it was possible to ensure approximately proportionally balanced coverage of all ecological units. This deliberate weighting process is known as stratification, but rarely is it applied in the manner adopted by Foley. However, purely random selection of sampling points, without the intervening segmentation, leads to uneven representation of critical landscape characteristics, so undermining any attempt at meaningful statistical inference. In Owens Valley, where again it was hoped to discern significant variation of behaviour from one ecological unit, or fabric, to another, Bettinger simply allotted a proportional number of sampling points to each unit (*stratum*). Within a 42 x 27km transect, he surveyed "the entire surface" of ninety-five 500 x 500m tracts (*segments*) selected from a possible 3,290 (Bettinger 1977, 7). The greatest disadvantage of this method is that it still allows the possibility of untestable bias within the ecological units. By taking samples from every segment Foley not only enhances comparison of one ecozone with another, but is able to say with some confidence whether or not there is significant variation within an ecozone.

The greatest potential problem with Foley's method is that the mean sample size from the sampling points might not be large enough to allow the application of a full range of analyses. At each point an area of 1000sqm was sampled, compared with a surface area of 25000sqm in one of Bettinger's tracts. Had there been a high incidence of artefacts this would not be a problem, but very few diagnostic items were recovered (for instance, there were only 92 scrapers, the most common tool, from the whole survey (Foley 1981, 188), making it impossible to refine chronology, and severely restricting interpretation of locale function. What Foley's method does allow is the forming of hypothetical assertions based on general models of population location and density, inferred from the distribution of artefacts. In more productive regions chronological and locale-dependent variations might have been discernible, conceivably enough for inter group hierarchies to have been inferred, such as those suggested from the South Dorset Ridgeway survey (*2.3iv*, above). But even with a wealth of finds the application of such a thinly dispersed set of random sampling points will fail to characterise sites (settlement, ritual or otherwise) in sufficient detail for an assessment of their relative status. Consequently there is unlikely to be any information about the structuring of authority, division of labour, or nature of ritual behaviour. Foley successfully demonstrated that certain very general classes of economic behaviour become visible, but the particular frequency of sampling was such that if the same plan was adopted in a setting where ancient agriculture was anticipated, even manure zones would go unnoticed. The only means for overcoming these problems, should a higher level of economic or political information be required, is to increase the frequency of sampling points. There would be a need for more labour, a smaller survey region, or a different manner of stratification.

For the Stilo regional survey, Calabria, Ian Hodder and Caroline Malone compared the amount of time invested in, and the quality of information collected from, an adaptation of Foley's "cross" with two other collection strategies. The first method, aptly described as "non-systematic judgement sampling", focused on likely find spots such as "ploughed fields, slight rises and areas near springs". A team of six covered 250,000sqm in a day. In the second experiment the same area was divided into four 300m wide bands. Within each band a parallel 30m wide transect was walked by six people abreast in lines at 5m intervals. While walking they paced out and recorded the length of each soil type and all areas that could not be searched within their lines. It was estimated that 79,410sqm were covered in one day. In effect this represented a 10% sample of the target area. The "cross" method comprised four 30m x 5m spokes, a total area of 600sqm per sample point. The same area upon which the first two methods were tested was divided into nine adjacent 300m x 300m segments, within each of which three sample points were chosen randomly, yielding a coverage of 1800sqm per segment, or 2% of the surface (a sampling frequency 40 times greater than that employed by Foley). 16,200sqm were covered in one day, representing around 20% of the area covered by transects and 6% of that covered non-systematically. The positive effect was to increase the overall sherd count by a factor of 5 against the transect survey, and by 33 against the non-systematic survey. These factors were reduced to 2.5 and 20 when only the prehistoric pottery was taken into account. It was also noted that whereas 6 pieces of obsidian were collected from the cross survey, only one occurred in the transect survey, and none in the non-systematic survey (Hodder & Malone 1984, 125-127).

This latter point is important; it demonstrates that it is not simply a case of variation in quantitative resolution, rather there is the qualitative problem that a whole class of information might be omitted by less intensive survey. This need for specific qualitative information, as well as periodically enhanced quantitative information, has to be measured against the goal of survey to characterise a region within a specified time, with limited resources. At Shapwick control over the quality of fieldwalking was effected by specifying that 10 minutes should be allotted to walking each 25m strip, with an assumed breadth of vision of 1.5m (Aston & Costen 1994, 15). In contrast, at South Cadbury it has been assumed that the quality of search over a specified time will vary according to the commitment, concentration and energy levels of the fieldwalker. Walkers are encouraged to focus on the particular strip of land by marking it on either side with 20m ropes set between markers. Teams of two move the ropes, and whilst one walks and collects the other records surface detail (soil type, conditions, incidence of stone etc. Figure 3.11). Preliminary results suggest that the stimulation of working in pairs improves concentration, and enables faster coverage of the ground.

Other surveys have attempted compromises by applying intensive techniques on all, or a sample of, sites recorded during extensive sampling. These secondary samplings have sometimes been carried out during a return visit after preliminary analysis, but very often they have been conducted immediately when a scatter of finds was adjudged to have become a concentration. In either case, a variety of different intensive collection or recording techniques were employed. In Berkshire it has been demonstrated that line intervals of 25m are likely to miss significant sites. An area of low density concentration of probably Early Neolithic flint artefacts, approximately 60 x 40m, was located by employing a mesh of 5m grids over an area of 2.5ha. The whole area produced 136 flakes, 7 cores and 18 tools; 46% of the flakes and 55% of the tools occurred within the concentration. This, of course, was a recovery rate swollen by coverage of the whole surface. When wide-spaced walking was conducted, at 25m intervals, a single scraper and two flakes were the only prehistoric artefact recovered from the whole 2.5ha (Gaffney & Tingle 1989, 56). This density would not have been sufficient to provide the slightest indication of the possibility that there might be a significant site. A more recent trend has been to reduce the sampled area, and to increase the sampling frequency. Instead of walking with team members set 25m or 50m apart, then using a more intensive technique when a "site" occurs, the Tarragona survey confined itself to four 1km wide transects, but sampled every available field within them at 5m intervals. In the Southern Argolid, where visibility was "for the most part very good" (Jameson et al. 1994, 219) against a background of thinly spread *maquis*, the intervals were flexible, varying from 15m to only 5m when visibility declined. It was reckoned that a "five-person team could cover approximately 1.0 km² in one six-day week" (Jameson et al. 1994, 223), unless the terrain became difficult or the team encountered fences.

The landmark survey in Boeotia was probably the first to employ lines at 5m intervals and to aim at total coverage of the land selected within a total area of 2580sq km, assuming that each walker had a field of vision of 2.5m to the left and right (Bintliff & Snodgrass 1985, 130-132). This very intensive survey was designed to illustrate the range of "site" sizes, so that the most economical interval could be implemented for diagnosing the smallest likely sites. This allowed an increase of interval between lines to 15m.

In the Mediterranean, Near and Middle East, where surface finds have often proved to be voluminous, it has been common practice to count the majority of finds and to collect only a representative or diagnostic sample. In Boeotia walkers carried "clicker" counters, and from 1980 they were asked to shout out what sort of artefact-type they had encountered. Apart from forming part of the overall record of surface material density the method allowed the team to make instant judgements when density increased. On this impressionistic basis a secondary method of systematic random sampling was employed over the area of perceived increased artefact density (see *3.3ixb*, below).

The Sergemes Valley project, building on previous work recording exposures of structural material, sought to establish a correspondence between surface finds and buildings. Here, too, counting devices were used, and during the first season a uniform fieldwalking strategy was employed, in which team members were spaced at 25m intervals, divided into 100m length collection units. It was hoped that by treating the landscape as a "continuum... a background on which to evaluate the exploitation of the landscape" would be created (Dietz et al. 1995, 125). Only "representative" finds were collected.

A major problem for some surveys can be the lack of diagnostic finds of the target period surviving on the surface. Periods during which durable raw materials were used extensively generated diagnostic artefacts which survive well (although chronological resolution may be poor). Metals rarely survive well on the surface, and their contemporary ceramics are usually the fieldwalker's target. However, friable prehistoric pottery rarely survives once exposed by the plough. In Britain, for instance, surface scatters of Late Bronze Age and Iron Age sherds are very rare. Occasionally it is possible to predict the location of Late Prehistoric activity on the basis of other finds. It has often been noted that sites of this period have been distinguished by surface concentrations of burnt flint. In East Hampshire 19 fields which produced 1 to 2 pieces of Late Prehistoric pottery, produced an average of 42 pieces of burnt flint, compared with an average of 15 pieces of flint in 223 fields from which no such pottery had been recovered, and an average of 29 pieces in fields where Romano-British pottery had been found (Shennan 1985, 75). However, such an analysis cannot be generalised to other regions, as it is likely to be contingent upon the relative availability of a raw material.

Ultimately there will remain problems with fieldwalking in many environments. Often it can only be used on cultivated land where the plough exposes, but ultimately destroys the archaeology. Whilst a weathered ploughed surface may provide very good visibility results will vary year-on-year, as experiments from several surveys have shown. At Maddle Farm, walking in lines at every 25m, the same team of experienced fieldwalkers walked the same field over three successive years and produced distinctly different results each time (Figure 3.12).

Three teams using a 1metre wide rope "corridor" for surface collection over a ploughed field. One person scours the twenty metre length of the corridor, while a second bags the finds and records the ground and light conditions, topography and other information (see figure 3.3, above).

Figure 3.11. Roped line fieldwalking, South Cadbury Environs Project

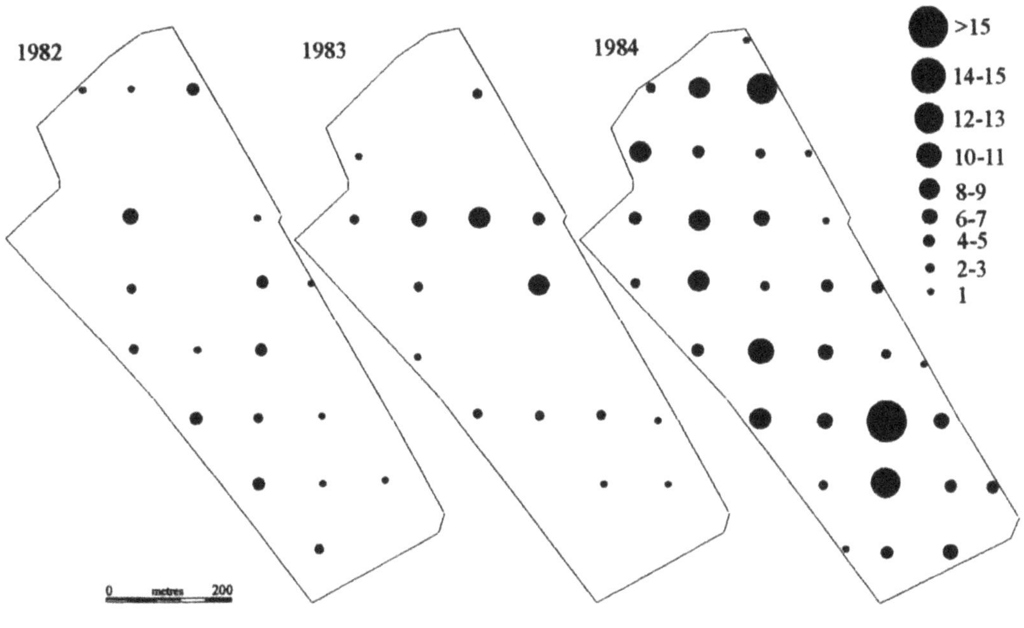

Figure 3.12. Flint distributions over successive years, Maddle Farm, Berkshire (from Gaffney & Tingle 1989, fig. 3.3)

3.3ixb   Fieldwalking: sampling the site

"Surface sites are foci both of activity and of habitation, and they are part of the spatial distribution of artefacts on a regional scale." (Jameson et al. 1994, 221)

The selection of areas of high density for more intensive sampling represents a compromise designed to increase general coverage of a region, while allowing enhanced chronological and functional resolution. In the Biferno Valley, when a "distinct artefact concentration was observed" the team made a decision about its extent, then "attempted to collect all the materials visible on the surface". When larger, classical sites were encountered, rather than sampling parts of their surface areas, the team applied a restricted collection policy which included all fine wares, a large representative sample of coarse wares, and a small representative sample of brick and tile (Barker 1995, 46).

Elsewhere sampling of sites has been more methodical. In the Southern Argolid, where a site was defined as "any location with ancient features such as architectural remains, or a concentration of materials ... having a recognizable boundary" (Jameson et al. 1994, 221; after Plog et al., 1978), a site "middle" was established, and its limits gauged according to when the artefact density declined to below one artefact per $10m^2$. If a site exceeded 1ha between 16 and 32 circles of $10m^2$ area were selected by computer and layout across the site as collection units (Jameson et al. 1994, 225). The penchant for circles was repeated on two sites investigated by the North Keos project. At Kephala 2m radius circles were set up at the 215 nodes generated by the meshing of 10 x 10m grids. All sherds and lithics within the circles were individually recorded, and "Exceptional pieces of chipped and ground stone were collected" for later study (Cherry et al. 1991, 202). At Paoura a less intensive procedure was used. A mesh of 25 x 25 m grids was laid out and at each of 271 nodes distinctive chipped stone was collected and other diagnostic material counted within 1.5m radius circles.

In Dalmatia a much simpler strategy was employed when increased artefact densities or structural remains were encountered. Team members would judge the limits of the scatter, then a single 5 x 5m quadrat was selected randomly, and all visible surface material collected from it. In an effort to sample 5-10% "of any given site" a further quadrat was added for every 20sqm by which a scatter exceeded 20sqm.

None of these methods is truly satisfactory. The first objection must be that impressionistic "encounter" strategies (i.e. diagnosis and sampling of sites during fieldwalking) are likely to be very subjective. More fundamental, though, is the failure to account for expansion, shrinkage and intra settlement shift over time. In some cases visible prehistoric remains have been shown to be very localised; although a settlement may remain in use over a long period, the structures within it, particularly if they were timber, may change location. Random sampling may very easily not only fail to note these shifts, but may detect no activity in periods during which the site remained occupied.

One of the most commonly applied methods for circumventing the percentage problem is "grab" sampling. This procedure requires team members to recognise and collect or record diagnostic artefacts, usually over the whole of the perceived site area. Although no claim is made for statistical validity, the assembled material is treated as representative of the chronological and, sometimes, functional range of a site.

A more sophisticated approach to grab sampling evolved in the course of the Boeotia survey, from 1979 to 1982. Initially grabs were collected to amplify chronological information about the "site". However, the selection and laying out of the random sample areas, which covered only 3 to 8% of a site, proved time consuming and in 1981 it was decided to allow full coverage of "small and medium-sized sites" (Bintliff & Snodgrass 1985, 133-134). For sites of up to 1ha this 100% coverage proved quicker than taking the time to select and layout areas for collection. Sites of more than 4ha were sampled by selecting a number of natural divisions, usually fields (which were very small in the region), which were sub-divided into 30 x 20m collection units for total coverage (Figure 3.13).

As justification for the selective collection strategy it was noted that the retention of 80-90% of 5000 pot and 500 tile fragments was symptomatic of "the success of the field-teams in selecting diagnostic material" (Bintliff & Snodgrass 1985, 134), an argument which is surely absurd. As work in the Tarragona hinterland has demonstrated (Carreté et al. 1995, 53) more detailed study of unpromising material can be very productive. More particularly, it has become clear that, for instance, the chronological span of some coarse wares at a site is not necessarily represented by easily datable, contemporary finewares. As Mattingly has stated "The key is that quantifiable assessment can be made of the relative density of different categories... of artefacts" (Mattingly 1992, 98); it is highly likely that there will be substantial chronological and functional biases in such a selective assemblage.

None of the methods given above are fully satisfactory. However, it is quite clear that, unless regions shrink to much smaller proportions to allow very intensive coverage, simple extensive techniques will not only miss smaller sites, but will fail to recognise those features which may distinguish the function and chronological range of one site from another. Without this level of discrimination any attempt at socioeconomic analysis will be severely hampered.

3.3ixc Ploughzone / topsoil sampling

Smith and Thorpe describe a technique, under the name "shoveltesting", as "the controlled examination of ploughsoil, whether newly ploughed or ploughed at some point in the past" (Aston & Gerrard 1995, 73). A pit is dug to a depth no greater than that to which the soil has been ploughed because characteristically it has been assumed that the soil medium of surface finds is identical with the homogenous product of ploughing, although replication tests have suggested that lateral displacement of surface artefacts is greater than for those within the ploughsoil matrix (Steinberg 1996, 370). If that assumption is made, the data provide a control sample by which surface visibility and fieldwalker's success rates may be judged; it may also be used to compare distribution and density of artefacts in areas where surface collection is never suitable.

In North America, during the late 1970s and through much

a) Specimen site sampling, 1979 method, Boeotia.
The extent of a site was calculated by recording the distance of the edge of an artefact density high at every 15° from a pole placed at the putative centre of the high. Up to 32 random samples were collected within the area from circular samples of 4 sqm.

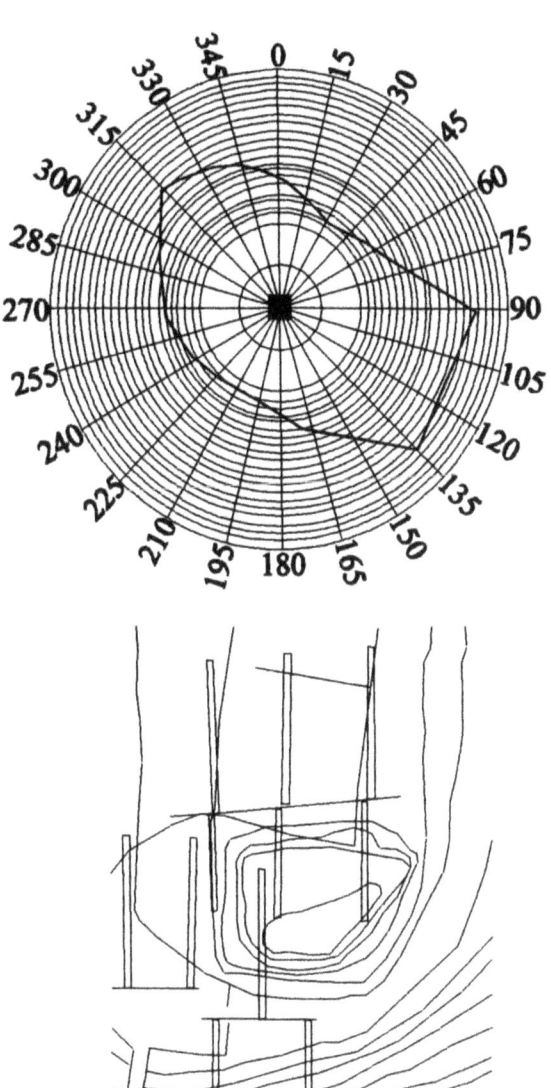

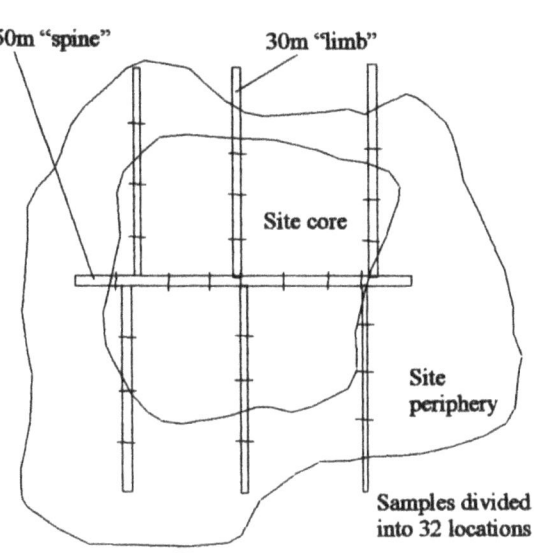

Schematic site sampling, 1980 method, Boeotia

Schematic site sampling, 1980 method, second stage, Boeotia

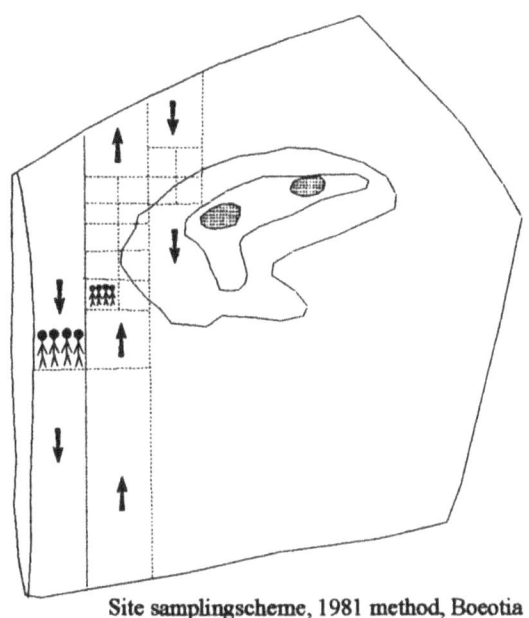

Site sampling scheme, 1981 method, Boeotia

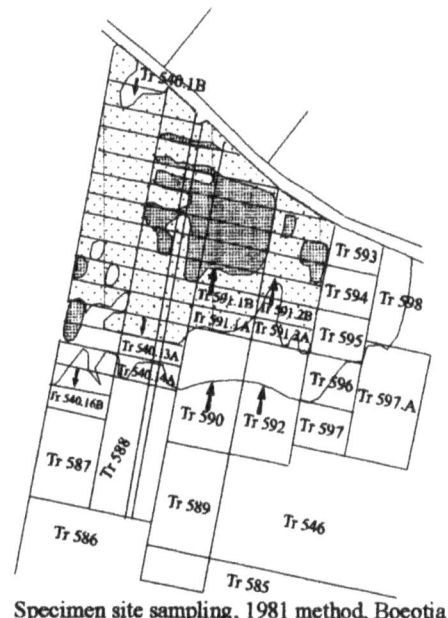

Specimen site sampling, 1981 method, Boeotia

Figure 3.13. An evolving site-sampling strategy, Boeotia. From Bintliff & Snodgrass 1985, figs. 4, 6, 7, 8, 9) of the 1980s, it evolved as a "discovery" technique, very often employed as a means for evaluating the archaeological potential of land due for development. As such its object was the "site", which came to be the focus of debate concerning appropriate sampling frequency and size. The arguments used here are equally pertinent to shovel testing sampling strategies, and will not be repeated in *3.3ixd*. Jack Nance and Bruce Ball considered that the "discovery probability" for a site was governed by "two independent variables": the intersection probability and the productivity probability. The former is the product of the relationships between four variables: the interval between and configuration of collection units, and the "size and shape of the surface area of sites" (Nance & Ball 1986, 460). Interval and site size were regarded as the most significant of these. Productivity probability is determined by the relationships between size of collection unit, the method of inspecting the soil from it, the "density of artefacts and the degree of spatial clustering of artefacts" (Nance & Ball 1986, 460). By using the concept of discovery probability Ken Lightfoot investigated means for using the technique as a truly regional tool by which probabilistic statements might be made, inferring the total population of sites from a limited sample (Lightfoot 1986, 490). By assuming a teleological perspective he invoked a simplified scheme, derived from Krakker et al (1983)[1]:

> "Theoretically, if the radius of a cluster of artefacts is greater than the midpoint of the square composed of any contiguous four test probes, then no matter how the site is placed in relation to the square grid, at least one test probe should pierce the site and lead to its discovery" (Lightfoot 1986, 492)

This might be regarded as slightly pessimistic, since the surface area taken up by each collection unit ("test probe") will increase the potential area of intersection of site and sample, an issue only dealt with at the final report stage (Lightfoot 1989, 414)[2]. The point of this is not to attempt to find every site, but rather to establish the probability of encountering sites of particular proportions. In theory, where a 10m interval is used between 30cm x 30cm collection units, every circular cluster of artefacts with a radius exceeding 7m should be discovered. It is possible to work out the probability of encountering sites of smaller radii using the equation $p=(pi)r^2/i^2$, where the probability of intersecting a site, p, equals *pi* multiplied by the radius squared, divided by the interval between collection units squared. Thus, to borrow the example used by Lightfoot, sites with a radius of 2.5m have a ".2 probability of being located", suggesting that for every one site found within that radius, it is expected that four would be missed (Lightfoot 1986, 497). In theory, the discovery of 3 such sites in a survey region implies a total of 15. If all site radii were processed within aim range, it would be possible to give an expected figure for the total number of sites in the area by adding the predicted small site totals to the number of large sites intersected.

Two problems must be tackled to allow this form of extrapolation. Firstly, how is the surveyor to know the area of a site observed during, for instance, a 10m interval survey?

At Long Island, New York, Lightfoot used the "iron cross" technique. Of the six stratified blocks selected for sampling, only one contained fields with a history of agricultural use, the form of which was not specified, so that by far the greater part of this survey cannot be described as ploughzone sampling. However, its methods are outlined here since it is doubtful that a shovel testing strategy would have employed a different approach. Where artefacts were encountered in a collection unit, secondary collection units were excavated at 2m from the primary collection unit in each of the north, east, south and west directions. If no further finds were recovered from these units the primary unit artefact was recorded as an isolated find. If further material was recovered from any of the secondary units the spoke in which it occurred was extended by excavation of another collection unit at a 2m interval (Lightfoot 1986, 494-495), and so on until a collection unit with no artefacts was encountered (Figure 3.14). Through this method, enhanced by some further adaptations for the final report derived from the plotting of artefact "fall-off patterns", it was possible to identify "areally limited sites with high artifact densities, spatially extensive artifact scatters with low artifact densities, and isolated finds" (Lightfoot 1989, 414). As an example of the range of distributed data, this ought to meet both off- and onsite regional survey criteria. The results compared well with a surface survey conducted by Lightfoot.

Michael Shott, in a somewhat hysterical response to Lightfoot, and to Nance and Ball, considered that the digging of small pits at regular intervals was a technique better suited to presenting general regional data, rather than attempting to locate and define sites (Shott 1989, 396). In particular, he felt that the use of a method such as the "iron cross" was likely to distort perceptions of site size; it failed to take into account that a site may have "blank" areas, so that treating a negative collection unit as a site boundary might be inappropriate. A simple, although more time-consuming, response to Shott's complaint might be to require at least two adjacent blank collection units along each of a cross's spokes before the boundary of a site was to be diagnosed. A more radical problem is that of detection. Shott rightly points out that whilst collection units may indeed intersect sites the evidence for the sites may not be perceived, whether from the ploughzone, shovel tests or archaeological test pits. This problem remains a topic of discussion (see *3.3xb*), but Lightfoot is surely right to assert that "unless one can demonstrate that the target population of cultural remains is detectable readily on the ground surface (an assumption often made but rarely demonstrated), then surveys methods should sample three-dimensional space" (Lightfoot 1989, 415).

Drawing on the North American literature, but informed by the example of Thy, Denmark, the Shapwick Project employs a different configuration of collection units by which five plough soil samples are dug in each 50 x 50m square, one at the centre of the square, and four at 15m in from each corner (Figure 3.15). Although much of the land was under pasture, the greater part of it was likely to have been under the plough at some point in the past, so that the soils might be treated

[1] Krakker, James; Shott, Michael J; Welch, P, 1983. Design and Evaluation of Shovel-Test Sampling in Regional Archaeology. Journal of Field Archaeology 10, 469-480]

[2] Lightfoot, K; Kalin, R; Moore, J; 1987: Prehistoric Hunter-Gatherers of Shelter Island, New York: An Archaeology Study of the Mashomack Preserve. Contributions of California Archaeological Research Facility]

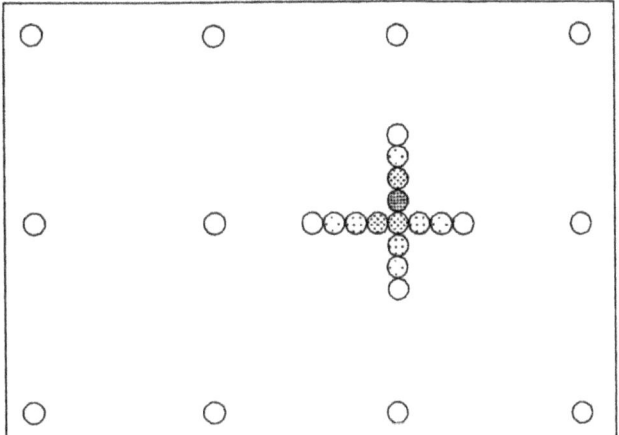

Figure 3.14. The "Iron Cross" sampling patter (as described in Lightfoot 1986, 494-95). Each circle represents a shovel pit. The darker the fill, the greater the number of finds. The wider interval between finds is 10m.

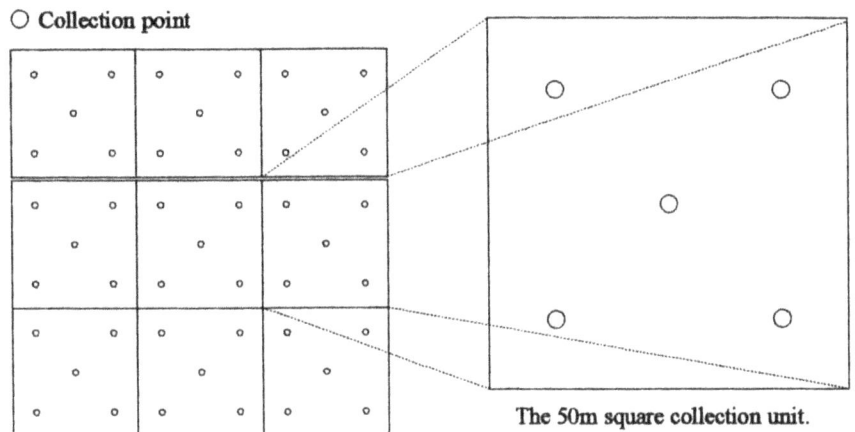

Figure 3.15. Shovel pitting sampling pattern, Shapwick, Somerset (as described in Aston & Gerrard 1995, 73). Each circle represents a shovel pit, five of which are set in each 50 x 50m square, within a mesh of contiguous squares.

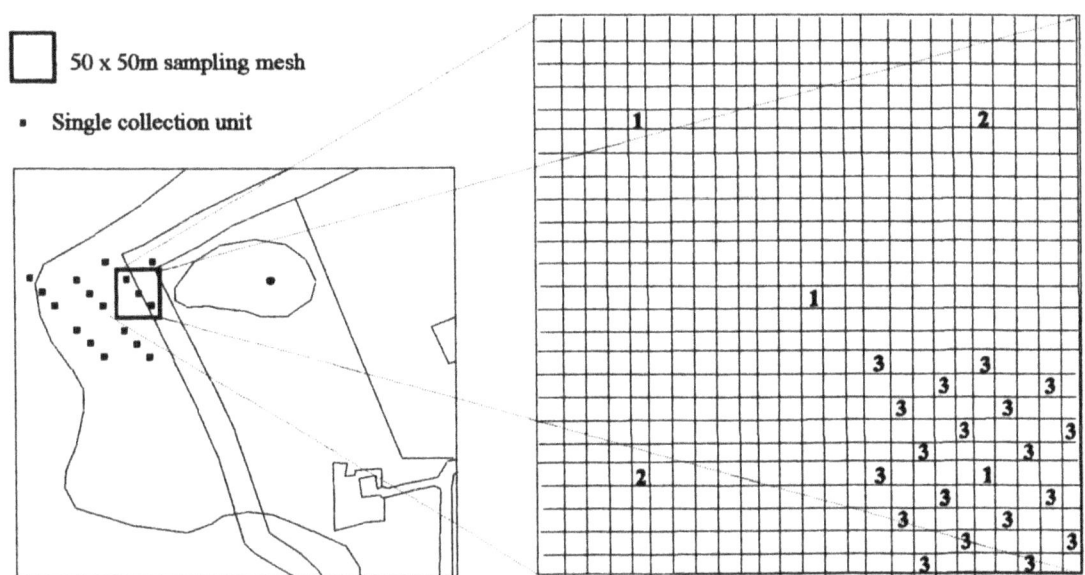

Figure 3.16. Offset-3 sampling pattern, Thy (after Steinberg 1996, *fig. 4*). Expanded view of a 50 x 50m mesh divided into 2 x 2m collection units. 1 - primary sample; 2 - secondary sample; 3 - tertiary sample.

as homogenous, and so described in volumetric as well as spatial terms. Each test comprised 30 litres of soil, which was sieved through a 3/8 inch mesh. The five tests were treated as a single sample which was assumed to represent a portion of the upper 20cm of soil, or 0.03% of 500m^3 (equated with 500,000 litres) of soil (Aston & Gerrard 1995, 73). The survey has yet to be fully published, so it is too early to analyse the efficacy of the technique.

Martin Millett has favoured the volumetric approach for the specific purpose of developing an effective means of comparing stratified rates of artefact recovery with those from the surface (Haselgrove et al. 1985, 31-37). But in work carried out at Thy, Jutland, Steinberg has pointed out that in cultivated landscapes the ploughzone may represent the greater part, even all, of the "site" (Steinberg 1996, 368). Quite simply, where the plough has repeatedly cut approximately 30cm into the ground it is common for occupation levels and the upper parts of cut features, as well as any relict superstructures, to be removed. If top soil is simply removed by machine, as is often the normal procedure prior to archaeological excavation, the homogenised cultural strata are lost, including by far the greater portion of artefacts which determine a site's signature (Steinberg 1996, 375; Schofield 1991b, 5. For a detailed treatment see *4.3iiid* below). Only the lower parts of post pits and gullies are left, and they may well contain but a very small proportion of the artefacts discarded on the site. Thus, the significance of the technique is altered from a means of fieldwalking control or enhancement, whereby sites are located, to the principal means of site investigation.

After allowing for homogenisation and wider dispersal of artefacts in the soil it was judged that intervals between samples exceeding 20m incurred the risk of failing to recognise sites, but by taking three samples along a diagonal line from blocks of 50 x 50m grids (an offset pattern. Figure 3.16) the number of samples required was reduced. The size of sample, by volume of soil, was determined by preliminary experiments to find the approximate range of artefact frequency. 100 litres per sample was considered to have been sufficient to represent the density of common artefacts, but a volume of 400 litres was preferred to facilitate the recovery of rarer lithic objects (Steinberg 1996, 371). 1000 litres of soil is equal to 1m^3 so that at the analytical stage it was possible to calculate artefact densities per cubic metre by multiplying the surface area from which the soil was recovered by the ploughsoil depth.

The project used a mechanical screener but even with this instrument it would have taken upto a day to cover 1ha (12 samples). When higher densities were encountered the sampling pattern was intensified, and upto 5 days were allowed on a "site". At this level of labour investment the method may be seen as quite separate from the variety of shovel testing adopted as a substitute or basic control for fieldwalking. However, it must be stressed that while the less intensive methods aim merely to represent underlying patterns (i.e., broken stratigraphy) the Thy method aims to describe the "site" itself (see *4.3iiid* below).

At South Cadbury variations on these themes have been applied, sampling each corner of a 20m grid (25 samples per hectare) but looking for greater resolution by treating each collection unit as a sample point, displayed within a GIS programme. The shovel test results are mapped onto the geophysical data plots forming the basis for subsequent interpretations as a part of a process of graphically based analysis. For more on this approach see *4.4iie* below.

In uncultivated topsoil lack of soil homogeneity suggests that artefacts may be distributed less generally, so that "sites" may appear significantly smaller, hence more easily missed. Density, as a proportion of originally discarded material, may well be greater than on comparable sites which have been ploughed. The implication must be that sampling intervals should be reduced.

3.3x   *Intrusive techniques*

Intrusive techniques are those which require the damaging of sealed archaeological strata. Augering rarely causes significant damage, but retrieves comparatively little information, only rarely including artefacts. Test pits are the smallest sampling units in which artefacts can reasonably be expected to occur in a statistically useful quantity, although contextual information may be limited or even misleading. The probability of causing significant damage to intact archaeological deposits will vary according to the size and frequency of the test pits.

Trial trenches and excavation are generally targeted on known archaeology and so are highly likely to cause damage, compensated for by full and accurate recording.

3.3xa   *Auger boring*

The classes of information recorded from an archaeological bore will be identical with that from a geomorphological bore. The difference lies in the sampling strategy. The latter is employed at a lower frequency, but the former is used at a comparatively high frequency. Its principal objects are cultural layers indicated either from excavated stratigraphy, from geophysical prospection, or from the incidence of surface artefact concentrations or sub-surface survey. In the Biferno Valley a simple screw auger proved effective in locating the extent of charcoal-rich layers which, although invisible from the surface, were considered to be associated with surface material (Barker 1995, 51). It was also hoped to "map outlying archaeological features around sites", and a sampling frequency of 1.5 to 2.5m (Barker 1995a, 83-90) makes this a reasonable aspiration - although interpretation from this technique will be highly unreliable.

At Brean Down their was little distinction to be made between geomorphological and archaeological objectives and analyses (Bell 1990, 90-94), but the use of a 10cm bucket auger allowed a fuller representation of the stratigraphy, supplementing vertically detailed but horizontally limited information with data showing the extent of activity during different phases.

Augering was applied with similar aims at midden sites in Kentucky, by boring manually with a sampling tube attached. Initially a right-angled cross pattern was sampled at 5m or 10m intervals, moving from the centre of the perceived midden. Further spokes were added and, at the ends of some spokes, where data indicated the probable presence of the site boundaries the sampling interval was reduced. At Carlston Annis over 30 person days were invested in an area of less than 1ha to obtain 97 cores of up to 2m depth, but evidence from elsewhere suggests that this was not a particularly efficient rate of progress. At the end of the work the extent and condition of the midden was well understood and the pre-midden topography was mapped (Stein 1986, 514-520).

Although the Brean Down and Carlston Annis examples are not taken from the kind of survey which is the object of this study, they provide useful examples of the sort of information obtainable through archaeological augering. It is clear that it is not a technique to be applied without foreknowledge that archaeological deposits are present. On the one hand, the samples per collection unit are so small that only concentrated cultural evidence is likely to provide a reliable indicator of activity and, on the other hand, any information is obtained at great cost in time. It may usefully be applied where an excavation, or a series of test pits, have established stratigraphic parameters which provide targets against which auger results may be measured.

3.3xb   Test pitting

Archaeological test pits may be used, like augering, to map the extent of a "site", but they are far more likely to cause damage, since a significant volume of cultural soils will be removed. However, test pitting may reasonably be used as an evaluative technique where development threatens an area with archaeological potential, or in a location where, for instance, woodland or deep soils render ploughzone techniques inappropriate.

Test pitting was an important strategy in archaeological evaluations carried out in Hampshire and Berkshire in the 1980s and early 1990s and formed the subject of a study by English Heritage. The technique's efficacy depended on three variables: the size and frequency of test pits, and their layout pattern (Champion et al. 1995, 53). Simulation tests were carried out using actual sites where there had been total topsoil removal, revealing subsoil features. It was found that an offset pattern, rather than a strict grid pattern, was the most efficient, but that there were greater difficulties in finding the balance between small, frequent pits, and fewer, larger pits. Although the wider dispersal of small units is more likely to encounter archaeological contexts, there was a strong chance that they might not be recognised. In larger pits, intersected contexts were usually visible when present, but the increased investment of labour time per pit will drastically reduce the total number of pits (Champion et al. 1995, 53, fig. 8).

The English Heritage study acknowledges that because its simulations were derived from positive evaluations, which subsequently were excavated, it has little to say about areas where no archaeology was discovered from the evaluation sample. There is a very high risk that a substantial number of "negative" assessments may be due to inadequate sampling rather than a lack of archaeological deposits within the target area.

However, the technique itself is open to criticism for other reasons, too. The minimum sampling frequency is determined ideologically, on the basis that there are particular classes of site, the dimensions of which are within a known range. While feasibility necessitates hard decisions concerning the amount of work that can be carried out on any project, there is a high probability of bias against the inclusion of small scale activity areas which may have had important functions within the regional economic scheme. Their use may also create chronological bias against periods where most activities are likely to have generated artefact concentrations only within a very narrow area. While this sort of loss may be acceptable for evaluations where development threatens the archaeology it is not so for a regional project which aspires to providing a full account of social economy. The arguments are closely kin to those rehearsed by Nance, Ball, Lightfoot and Shot; (see *3.3ixc* above).

On this evidence, in the regional context, test pitting may be useful in assessing the extent of a "site", but it is not likely to be an effective means for discovering one, unless there is another class of evidence which may help. Barker reports successful targeting of Early Medieval sites on "densely forested hills" in the Tuscania and other Italian surveys, on the basis of documentary research (Barker 1995b, 3), whilst at San Vincenzo, using similar evidence, a survey discovered "small, loosely nucleated settlement in secondary woodland or scrub" which would have been impossible by surface collection (Hayes 1985, 135). However, documentary or epigraphic data are only available within a limited range of periods and places!

The Maddle Farm project used test pits ("sub-surface survey") in a very different manner, specifically as a means for exploring the relationship between the surface and the soils down to bedrock. Blocks of 1ha were chosen from an area which had undergone extensive survey (lines spaced at 25m) and pits of 1 x 0.5m were excavated at every 25m along the fieldwalking lines, giving a total of 16 pits/ha (Gaffney & Tingle 1989, 23). High densities of surface pottery were reflected in sub-surface concentrations, excepting where stratigraphy had remained intact, when the sub-surface incidence rate was disproportionately higher than that on the surface (Gaffney & Tingle 1989, 24). The South Cadbury Environs Project's programme of digging to natural one 1 x 1m pit per hectare provided a clearer understanding of positive and negative surface artefact data (Figure 3.17), while also helping in the assessment of the chronology and even function of geophysical anomalies. The pits also offer a history of soil movement in the sampled areas.

The prospects for archaeological test pitting within a regional survey strategy are worse on deep soils; Crowther and French have pointed out that under current budgetary constraints "those questions which stress the continuity of the archaeological record... are not appropriate ones to ask of an entirely buried landscape" (Haselgrove et al. 1985, 67). Despite this, an effective programme combining augering and pitting was carried out at Seamer Carr, Vale of Pickering. The project's specific aim was to locate concentrations of artefacts at Mesolithic levels in environments which would have been similar to those at Starr Carr. A first stage comprised over 100 contour transects, with augering at 15m intervals. From the consequent information eighty 2 x 2m pits were dug at points which were within 2m above the target strata at 25m OD. Some flint was recovered from over half of the pits, six producing in excess of 10 pieces per 1m2. Eventually work focused on one area over 1000m^2 of which were dug to the 25m OD level over six seasons (Schadla Hall et al. in Haselgrove et al. 1985, 82).

The Seamer Carr method is not likely to form an extensive part of a regional strategy, but the principal of combining

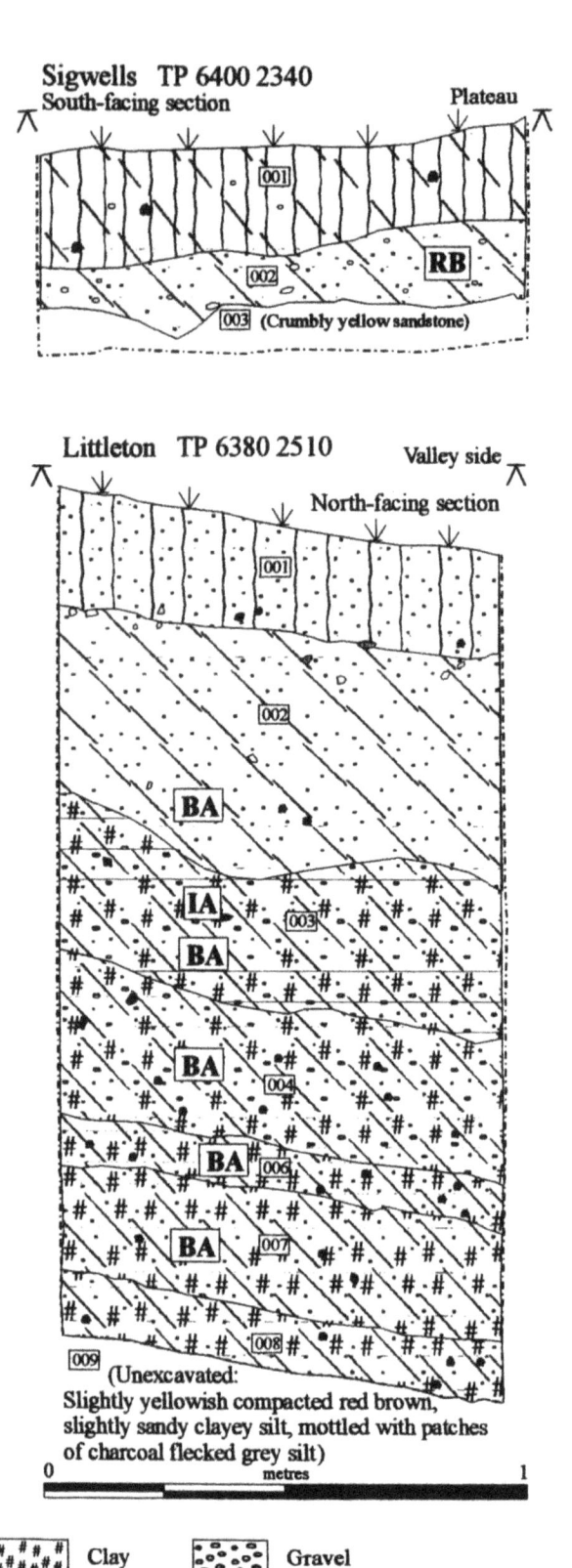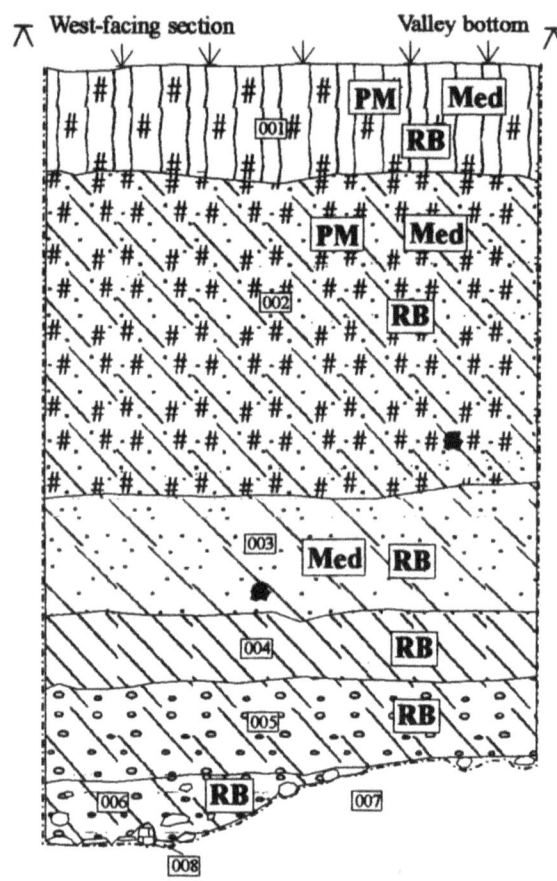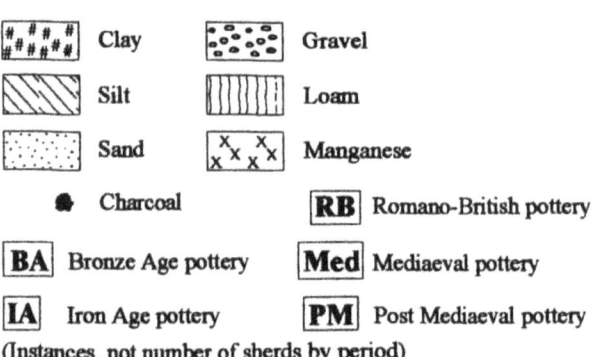

Figure 3.17. Test pits from various topographic zones, South Cadbury Environs Project

test pitting with auger survey is apt. Augering should be preceded by geomorphological test pitting, and Schadla Hall et al. demonstrated that it ought to be followed by archaeological test pitting.

3.3xc   Trial trenching

Archaeological trial trenching is here designated to mean the cutting of a section through or across an area where the results of other techniques suggest the likelihood of archaeological deposits. Typically trenches will sample air photographic features, geophysical anomalies and areas where concentrations of surface, ploughzone or test pit finds have been recovered. The trial trench may be distinguished from the test pit because it addresses a specified target, and is not used as a means for sampling an area where there is no evidence for archaeological deposits. It is likely to be larger than a test pit, and because it is located in response to particular archaeological data it may function in isolation, or as one of several with no formal sampling pattern. The trial trench may be distinguished from an excavation by its limited goals, principal among which are likely to be the acquisition of dating evidence, the nature of stratigraphy and quality of preservation.

Martin Bell has used trenches of upto 30m long across valley bottoms on the South Downs, the length of which were exposed down to natural before a 1m wide trench- length was excavated by hand (Bell 1983). Although the first aim was to investigate the character of valley sedimentation he was also able to demonstrate probable Early Bronze occupation sometimes below more than two metres of colluvium, alerting survey archaeologists to the potential inaccessibility of targeted data.

3.3xd Excavation

Increasingly, excavation is recognised as a desirable survey tool. John Cherry has implied that there will come a time when survey will supplant excavation as the dominant practice in field archaeology, but he goes on to say that "In principle it is hard to argue against the notion that regional surveys should include excavation as an integral and essential part of their strategy" (Cherry 1994, 94 & 103). It remains the most labour intensive, expensive archaeological technique and can only be applied to meet specified, limited, objectives often concerned with the need for stratified type series. In the Tarragona hinterland survey, in contrast to many recently published surveys, great emphasis was placed on the importance of coarse ceramic wares; by reference to well understood stratified deposits excavated at Tarragona, and to excavated material from neighbouring regions, it was possible to build a full type-series which was applied to the material recovered from the surface survey (Carreté et al. 1995, 63).

The Africa Proconsularis project sited 11 trenches, the surface area of each being less than 10sqm (a scale of work which might better be presented as trial trenching), in the area surrounding their target town of Sergemes, Tunisia. Although the trenches were placed deliberately over a diverse selection of stone structures recorded during field survey "it was never the intention to identify the function of the structures. It was hoped that the ceramic material could constitute the basis for an evaluation of the survey method and also confirm, or disprove, whether the dwellings in the countryside and the town of Sergemes were contemporary" (Dietz et al. 1995, 383). To this end stratigraphic integrity was emphasised.

The comparison of surface material to a stratified sequence is entirely appropriate, but more ambitious attempts to characterise period specific activity within a surveyed region are likely to produce unfounded interpretations. Barker felt free to calculate that if 200 sherds were collected from a 30 x 50m surface at Fonte Maggio, and some 6000 sherds were recovered from an excavation with a surface are of 30sqm, the whole site would produce around 180,000 sherds (Barker 1995, 140). His more modest assumption that, since four lithic concentrations associated with a few Neolithic sherds proved to be settlements of that period when excavated, all lithic concentrations accompanied by Neolithic pottery overlie settlements has the ring of common sense about it, even if it lacks theoretical rigour (Barker 1995,103). For Cranborne Chase it has been suggested that largely unsystematic fieldwalking recovered the full range of flint artefact types in proportions which were repeated in excavation; the durability of lithic material makes this claim more likely (Barrett et al. 1991, 28). However, work elsewhere has demonstrated that where a detailed horizontal record is kept it is normal for there to be sharp fluctuations in the densities of finds across a large excavated area.

The proportion of surface to stratified finds at Kephala, Keos, was used to estimate the volume of erosion. The near 1:2 ratio of both less durable ceramic material and durable obsidian was treated as an indication that approximately 50% of the "archaeological sediments may have been removed" (Cherry et al. 1991, 205). Similar work would have to be carried out in many other areas before the validity of such an assertion could be evaluated.

Although test pitting, trial trenching, exposure sampling and even augering may provide environmental data, almost invariably major reconstructions of landscape vegetation, or the faunal record, rely upon excavated material. Most of the evidence for woodland cover and clearance in Neolithic and Bronze Age Cranbome Chase is derived from two ditch sections (Roy Entwhistle and Mark Bowden, Barrett et al. 1991a, 20-48). In the Biferno Valley survey heavy reliance was placed upon the floral and faunal assemblages from Fonte Maggio (around 550 carbonised grains, beans etc.) and Masseria Mammarella (less than 100 carbonised grains), although cores from sediments were a further source of data.

The use of excavation as the primary technique for fieldwork on the Danebury landscape has enabled the collection of a large body of ecological information (Cunliffe 2000, 39-74). In addition to bone and carbonised flora from *sites*, a further survey using augering and pits in valley bottoms close to the hillfort collected mollusca and investigated the distribution and date of sediments using radiocarbon and thermoluminescence (Williams and Evans in Cunliffe 2000, 39-43). Whilst far from comprehensive it represents and far richer and more widely representative environmental data resource than has been generated by other surveys.

Ultimately, issues such as settlement structure, local and intra-regional exchange, and chronology are only likely to be addressed effectively with some excavation. At least a substantial programme of archaeological test pitting is likely to be necessary to obtain the level of environmental data necessary to build detailed accounts of subsistence and economy.

3.4 Spoilt for choice

The panoply of techniques outlined and discussed above are by no means exhaustive, but they give an indication of the range that is specific to, or can be adapted for, regional survey. Whereas until the early 1980s most surveys relied upon three or four of these techniques (typically, mapping, survey of the literature, fieldwalking) the modern survey is likely to apply a dozen or more of them. Quite often use of augering or geophysical prospection have not been applied in a truly regional manner, usually being targeted on a few presumed "sites", but gradually the potential for the latter to provide information about the division of the landscape is being recognised. It can also be used more widely, but requires complementary use of more intensive methods such as test pitting.

The physical characteristics of a region should be a key factor in the choice of techniques but, before decisions are made, project directors must be fully aware of the type of information each technique will provide, and whether or not he or she has access to labour skilled in its execution both in the field and for subsequent data analysis.

It is only by reading the expanding survey literature and through experience that practitioners will be able to judge which techniques are most likely to provide data commensurate with the issues they are addressing in a particular region. The direct importation of collection strategies applied in semi-arid environments into temperate zones will usually cause waste of effort and misleading results. It is essential to have a clear understanding of why a method should be adopted, how it may be used, and to find means of control against which the data collected may be assessed subsequently. Means of control can only be established by intensive, usually invasive procedures, and they will command more resources. However, they should be employed at every locality within a survey where variations in the land form, whether natural or due to human agency, may have a significant distorting effect upon the pattern of data distribution.

4: Time, space and the thing

In Chapter 2, I presented a simple scheme (Table 2.1) outlining narrative progression from raw data, through data processing to accounts of social structures. That scheme sought merely to describe the categories of narrative which people are likely to employ; although intuition may suggest that interpretation develops from the basic to the complex, the principal aim was to exemplify the sorts of narratives which surveys have produced, not to explain their production. In this chapter I shall attempt the explanation, showing the relationships between data, data collection techniques and strategies, and the intermediate interpretative phases which precede the full-blown narrative.

In some respects the issues concerning methodological rigour and the significance of the data arose from the work of Lewis Binford and David Clarke, who pressed archaeologists to ask questions not only about the data, but about the manner of its collection and presentation. If one is to apply an ethnographic analogy to a culture the limits of a particular excavation, even when set alongside several others, tend to offer a biased impression of the material record, reflecting only a partial view of the system being studied. Ethnographers had been observing activities carried out by members of groups over much wider areas and the activities varied in different parts of those areas. While excavation may reveal some of the wires entering the Black Box (Clarke's analogous representation of a complex system visible only by its input and output terminals; Clarke 1978, 58-62), across a region there will be many more lengths of cable which will give us more opportunity to make inferences about the mechanism inside. However, an arbitrarily chosen region for investigation should not be conflated with any previously existing socio-cultural system or systems which may have existed within its limits.

Although there is more widespread agreement that the artefact is the fundamental unit of archaeology (Foley 1981, Cherry et al. 1996), whether excavated or recovered during survey, a term such as culture is often criticised for introducing preconceived ideas of structure, when represented as the system within which the particular object acquires significance. Artefact-defined cultures readily assume the status of entity, long before we are able to judge the sense of affiliation experienced by the individuals and groups who form the objects of archaeological research. Against this is the view that unless we look for regularities within the data from which categories can be drawn, archaeologists are glorified treasure-hunters.

Binford has asserted that "We must have... common points of reference between our experience and the 'past', if we are to reason to the past in any realistic sense" (Binford 1982, 59) and has made use of detailed ethnographic data to establish those common points when reconstructing sites and the social structures within them. Although making use of ethnographic data to reinforce his analytic scheme, Clarke believed that the archaeologist ought to start with a category based system, classifying the definable attributes which distinguish varieties of artefacts, presenting them in *Iconic* ("distribution maps, histograms, scatter diagrams"; Clarke 1978, 31) or, preferably, *Symbolic* form. By the latter, attributes are "integrated into a specific calculus.... of a deterministic, statistical or stochastic kind such that reality is approximated by the logical mathematical consequences of the model" (Clarke 1978, 33). This was an approach which inspired the analytical methods of several ambitious young researchers two decades ago, but it is even more telling to compare reports for two surveys recently published under the editor/authorship of Graeme Barker. When creating a narrative based on work conducted from 1974-78 in the Biferno Valley (Barker 1995; Barker 1995a) Barker relied upon a survey-generated gazetteer, his knowledge of comparable work, his archaeological imagination and his skill at presenting an argument in the written word. The result is a very satisfying edifice resting on insecure foundations. His work in the Libyan valleys from 1979-89 (Barker 1997) has also resulted in a compelling narrative - but this time it is supported by bar charts (environmental data, ceramics) and statistical analysis (the distribution of walls).

The particular character of archaeological data forms and materials governs the quality of inference which may be derived from them. In the first place, we have to bear in mind that data are not absolute facts, but conceived entities; objects of perception. The contexts in which they are perceived vary with the methods of collection and recording. The following section considers the particular significance of data being collected within a region.

4.1 The perceived

"I suddenly see the solution to a picture-puzzle. Before, there were branches there; now there is a human shape. My visual impression has changed and now I recognize that it has not only shape and colour but also a quite particular 'organization'." (Wittgenstein 1958, 196e)

In the field, once the parameters of the area to be worked in have been measured, the first task that the archaeologist must be able to carry out is the class A diagnosis that he/she has come upon an artefact. He/ she must locate it in space and later establish its chronological and functional characteristics. However, if we accept Clarke's dictum that an artefact is anything fashioned by human agency it quickly becomes clear that in reality even the archaeologist practising total collection within a given area leaves much more than he/she keeps. This is particularly and most obviously true when we are working in three dimensions where as a matter of deliberate policy a great volume of material which has information potential is shovelled into a bucket before being discarded on a spoil heap. A section drawing represents a plane of perception; however, the reliability of the information it represents declines rapidly, since what is conveyed concerns a specific portion of space. The drawing condenses components of various strata, each of which may themselves comprise some consequences of processes which have had a duration ranging from seconds to centuries. However, a section drawing is not merely an instance of stylised compression; it is the product of perception, selection and exclusion.

Clarke made a virtue of selection; while recognising that modern perception of an ancient artefact is very different from that of the culture which produced it, he had confidence that archaeologists would be able to agree on not only artefact types, but the particular attributes which enable us to classify and compare them. Archaeologists would not necessarily focus on the same classes of attributes because different research objectives would call for new criteria. Bradley appears more sceptical, maintaining that we can only see what "we can imagine as possibilities" (Bradley 1987, 117).

There is a fine line between recognition by analogy and imagination; both these means for designating a thing rely on the experience of the perceiver. Invoking "imagination" seems merely a ploy to avoid accounting for how that designation was arrived at. If we describe a sherd as part of a bowl, imagination is a lazy term which avoids accounting for a skill derived from experiencing similar shapes. There are constraints which psychology has demonstrated will limit our range of descriptive terms, represented by two basic variables: the curvature of the bowl/non-bowl and relative proportions of the population which may be expected to see it as a bowl/non-bowl. But the chosen term is not based merely on an abstracted geometry; regardless of the origin of the word its range of meanings will fluctuate according to criteria other than its shape. In the modem kitchen the term bowl can refer to flat-bottomed objects which appear subrectilinear in plan view; the washing-up bowl. Thus function can be at least as important a determinant as shape.

Famously, Wittgenstein employed the duck-rabbit (Wittgenstein 1958, 194-206) to consider the contextually sensitive ways in which a deliberately ambiguous line drawing might be perceived. His example was derived from an image used by an experimental psychologist, constructed for the purpose of research into perceptual ambiguity. The ambiguity we experience when confronted with a broken object is that we can neither be certain of the intentions of the once complete artefact's producer, nor do we have analogous experience of its use. We see ducks, and we see rabbits, situated within our chronological dimension (admittedly a fuzzy concept), and although there remains a debate amongst psychologists and linguists (McAndrews et al. 1997, 82-109) about the logical problems associated with memory by image or verbal concept, we routinely experience the ability to recognise classes and sub classes of thing. The chronological dislocation of an ancient artefact from our present experience does represent a problem because we do not have direct experience of different members of a class of object in its temporal, social context; nonetheless, in most cases it is that experience upon which we rely.

John Barrett, discussing the Bronze Age pottery of Cranbome Chase, claimed to have dispensed with the traditional categorisation of artefacts. He presented a view of society through that time as characterised by emergent differentiation of human agents, who are "aligned towards certain forms of social authority" by way of "the practical, metaphorical association of people with a certain range of culturally specific values". In this process:

> "Material culture represented a grid of material categories which became the 'props' or reference points enabling the agents to re-situate themselves and others in a complex matrix of metaphorical associations." (John Barrett in Barrett et al. 1991a, 205).

The important, if simple, point is that the way in which we categorise the world is based on our experience. We also categorise the prehistoric world according to our experience, not that of those who dwelt in and conceived that world. Barrett draws on psychological perspectives of categorization explored by Eleanor Rosch. Published extensively in the 1970s, she based her analyses on experimental work with people of different language groups, ranging from North American, English-speaking university students to adults of the Dani of New Guinea. The Dani were chosen because of the extremely restricted range of terms for colours. Rosch's explicit aim, using a wider range of focal and non-focal colours from Munsell, was to disprove the "language cognition hypothesis" which had been propagated variously, first by Edward Sapir and latterly by Benjamin Lee Whorf, as "linguistic relativism" or, more strongly, as "linguistic determinism". The strongest form states that language, syntactically and semantically, determines how we think.

Rosch felt that she had successfully undermined the hypothesis (Rosch Heider 1972, 19), although evidence from other work compelled her to plead "special case" against results from experimental work carried out by P. M. Greenfield and C. Childs[1] which "studied the effect of knowing certain patterns in cloth upon pattern conception". Experimental subjects were classified in two groups, both of which "could discriminate differences between red, orange, pink and white" but the members of only one group could name the full range of distinctions, while the members of the other group had names for red and white only. Rosch gave the following account of the experiment:

> "The patterns consisted of simple groups of red and white threads. Subjects were asked to "copy" the pattern by placing a stick into a frame. They were given their choice of various widths and colors of sticks. While some subjects used only the red and white sticks to copy the red and white patterns, others freely substituted pink for white and orange for red.... The important point for our argument is that it tended to be the subjects who named the red, pink, orange, and white sticks with different names who adhered strictly to the red and white sticks for copying patterns; subjects who used only a single term for white and pink and a single term for red and orange were the ones who tended to make the substitutions." (Rosch 1974, 116)

Rosch recognised that this result suggested that language was a significant determinant of behaviour in this experiment, but pleaded that: "It may well be that it is in the little understood domains such as aesthetic judgement that the use of color terms will be found to "make a difference". (Of course, the Zinancantecos who used differentiating terms may have done so because they were more sensitive to aesthetic differences)" (Rosch 1974, 116). Significantly, Greenfield and Childs chose an experimental method in the form of a process which may have had some kinship with life experiences of the subjects. Rosch's rigorous methods created a more rarefied atmosphere, a western scientific scenario, which may itself have influenced the subjects. It

[1 Greenfield, PM and Childs, C. 1971. Weaving skill, color terms, and pattern representation among the Zinancantecos of Southern Mexico]

might appear that Rosch, caught up in a world and language of empiricism, had to create a separate universe, one of aesthetics, to which deviations from her findings can be consigned. What Rosch wanted to find were "natural categories" which transcend the contingencies of language.

Rosch observed a tiered taxonomy in language (Figure 4.1) which included superordinate, basic and subordinate categories. Members of superordinate categories are abstract and comprise members which "share only a few attributes among each other", whereas members of subordinate categories share many of their attributes with members of other categories from the same "basic" family. Thus a kitchen chair and armchair will each be inanimate, have four legs, a back support and a platform for sitting on. The concept of *cue validity* has been used to described the probability that a thing belongs to a particular category. In the case of a chair, "having four legs" is one of the cues which increases the likelihood of a thing being a chair. Thus the reliability of this particular cue can be represented as a probability by dividing the number of members of the universe of chairs by the number of members of the universe of things "having four legs". It would quickly emerge that a very substantial number of things that are not chairs do have four legs, so that this particular cue is a poor predictor of category membership (Rosch et al. 1976, 384-385). Of course, the efficiency of a cue can depend on its medium. If the cue "having four legs" is conveyed in a simple drawing, the construction of the image will significantly influence the subject. If the person doing the drawing which is to be presented to the subject knows that the cue is for a chair, unless he/she sets out to deceive, the legs are likely to appear inanimate. In contrast, if the draughtsperson is an intermediary without the knowledge that the cue pertains to chairs, he/she is just as likely to draw the four legs of an animal.

By the mid-1970s Rosch had moved if not to a linguistic determinist position, then at least to a qualified relativist one, recognising that

"Basic objects for an individual, subculture, or culture must result from an interaction between the potential structure provided by the world and the particular emphases and state of knowledge of the people who are categorizing it. However, the environment places constraints on categorizations. Human knowledge cannot provide correlational structure where there is none. Humans can only ignore or exaggerate correlational structures." (Rosch et al. 1976, 430)

Whilst it is surely true that the environment exercises constraint over categorization, we may be a along way from a time when those constraints become apparent. This is particularly true where issues of religious belief, social economy and aesthetics are concerned. In these cases it is not simply that "categories... can have differing sets of correlational structures, depending upon the degree of knowledge of the perceiver" (Rosch et al. 1976, 430); a person can be fully acquainted with the economic categories proposed by Ricardo, but adhere to the tenets of Marx. That adherence may be the product of careful socio-economic analysis - or may be a means of expressing a dissatisfaction with authority which developed in childhood. This example might seem more pertinent to a discussion of higher levels of narrative; the problem is that previous and existing higher levels inform our categorising of artefacts.

Despite this Rosch's taxonomy can form a useful basis for artefact analysis. From the field archaeologist's point of view a particular set of taxa might comprise the superordinate "pot", the subordinate "Wessex/ Mid-Rhine", and the basic category "Beaker". To a ceramic specialist, "Beaker" might assume the place of superordinate, "tripartite, zoned, toothcomb impressed decoration" the subordinate, and "Wessex/Mid-Rhine" would become the basic category. Bearing in mind that the archaeologist is usually extrapolating from a fragment of a vessel, it is quite probable that the modern non-archaeologist (or even undergraduate student of archaeology!) might not even recognise it as a pot - or even "find"! However, if we can discover the dynamic patterns of our own varied modes of basic categorisation we may find a window into the varied taxonomies of the past. If hierarchy and specialisation were a feature of a society we may perceive patterns of constraint which vary according to the status and function of the individual or group. It is exactly

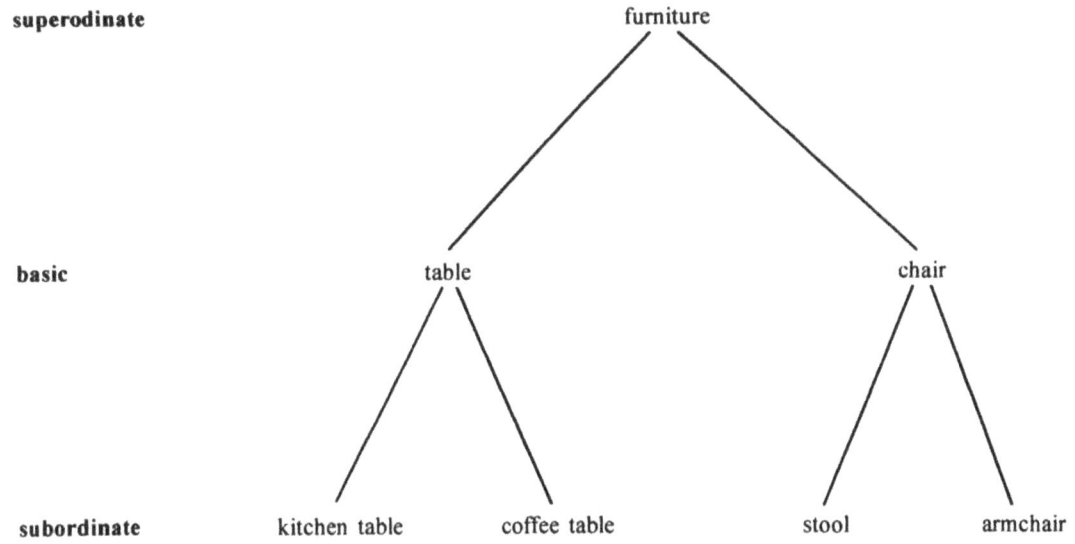

Figure 4.1. Language taxonomy (from Rosch et al. 1976, 384-85)

at this point that we may hope to use functional and distributional accounts to inform more ambitious levels of narrative.

If we are to develop a detailed understanding of the function of a particular area of occupation we need to have as full an account as possible of the use and significance of each artefact class and type, including those classes and types which occur rarely. Are there needs for such rigour in regional survey?

The answer depends on the aims and strategies of each particular survey. The ideal objective must be the collection of all portable artefacts, coupled with a level of class/type knowledge and analysis commensurate with (and inevitably based upon) that from excavation. However, there are few surveys which adopt such a practice, which tends to generate a wealth of post medieval material, itself expensive to store, process and analyse. This raises an important issue: is the archaeologist who collects recent material ethically obliged to conserve and analyse it? If he/she does neither, then information will be irretrievable for future field workers - as well as those already interested in the recent past. All artefacts were collected during fieldwalking in the Tarragona hinterland, but the post medieval material was sorted and discarded without record (Carreté et al. 1995, 52), a practice anathema to the Shapwick project. There, the collection of all artefacts has been preferred on the grounds that it allows "voluntary labour of varying expertise to be used and ensures that a range of material from all periods is represented" (Christopher Gerrard in Aston & Gerrard 1995, 10). The literature fails to demonstrate that either is wholly true, and marked disadvantages concerning storage, conservation and medium term labelling are described with laudable frankness (Aston & Gerrard 1995, 11).

The vast majority of projects have not attempted such a broad collection strategy; however, the collection and unrecorded disposal of material because it is not of the target periods should be regarded as unethical. The medical equivalent would be a cancer specialist striving to destroy that disease in his/her patient, but preventing the cardiovascular consultant from treating a life-threatening deep vein thrombosis. It is acceptable to leave a particular problem or period to another specialist, as long as any information gained by accident (a clot noted during radiography prompted by the cancer; a flower pot sherd collected because it was mistaken for Samian) is properly recorded; if surveys where not allowed to target periods many would be too expensive to take place at all.

Having accepted that it is legitimate to apply labels to classes and types of artefactual data, and having resigned ourselves to a policy of discriminatory collection coupled with the full recording of accidentally collected objects significant for other periods, it is important to have clear criteria for targeting finds. All surveys which are carried out with narrative intent - in other words, surveys which aspire to more than plotting the distribution of "sites" without regard for their spatio-temporal relationships to one another, must decide on their desired level of spatio-temporal resolution. In some cases, where the evidence is rare, the appropriate method of survey may be a review of literature from a few well recorded excavated sites from a large area. An extreme example is a survey of Late Palaeolithic settlement patterns in North West Europe, looking for evidence of subsistence strategies and group affiliation (Newell & Constandse-Westermann 1996). The rarity of the evidence was, in this case, due to chronological remoteness. Equally, a particular class of data may not be susceptible to surface collection; once again a regional perspective will be best served by the presentation and analysis of results from excavations in the region; a good example is the survey of Iron Age crop data in northeast England (van der Veen 1992).

But my chief concern is with those surveys in which data are collected from the field. What class of narrative (and for which periods) does a survey hope to achieve? What is the targeted chronological span, and what artefacts are associated with that span in the particular region? What types of artefacts are likely to have survived on the surface, and which of these will best illustrate activity, date and affiliation? Under what circumstances may an artefact be recorded, but not collected, because all significant information can be garnered in the field?

In practice the chronological span usually determines the range of substances of which the artefacts are composed. In very crude terms the principal finds prior to 1500BP in the Americas, prior to 2000BP in North Europe, and prior to 4000-3000BP in the east Mediterranean, Near East, and parts of Indian Sub Continent and China are likely to be lithic; after these dates they will be ceramic. Although lithic material always overlaps with the introduction of ceramics, the earlier ceramics lack durability. Over much of North Europe this is a particular problem since there are no proper lithic typologies for the Later Bronze Age and Iron Age (apart from larger objects such as querns), the pottery of which has a short life once exposed on the surface. In contrast, there is a substantial overlap of durable pottery with obsidian and other lithics in the east Mediterranean.

Other materials, such as glass and metals, have a poor survival rate on the surface, although waste from their production, in the shape of slag, lasts very well. However, slag, like bone (although only bones from domesticated or hunted animals may be considered as artefacts, from a stratified context "non-artefact" bones will provide valuable environmental data), cannot be dated by means other than association with other finds. Undatable classes of artefact are rarely kept, except on grounds of rarity or aesthetics! Ultimately we cannot escape a Clarke-like recognition leading to the analysis of characteristic attributes of artefacts, so as to set them within a temporal framework, prior to situating them in space. The different properties of artefact raw materials, in particular there durability, has a more dramatic impact on surface data than that recovered by excavation. It is appropriate to identify the effects some of these properties will have on distribution maps!

4.1i   Lithics

Although the word *lithic* pertains to stone in general a reader of archaeological reports, survey or excavation, might gain the distinct impression that lithic was synonymous with "chipped stone". The historical reasons for this selective approach are that: 1) stone is considered to be a weaker chronological indicator than pottery, and only assumes importance in the absence of it; 2) flaked or chipped stone carries the most highly resolved chronological information, which has been structured in typological sequences; 3) flaked or chipped stone is more portable, and also more likely to be recognised as significant by an inexperienced . Curtis

Runnels has written in polemical style in a cry for an inclusive approach. Worked stone can offer information about hand tools, larger scale industrial processes, personal adornment and architecture (Runnels 1994, 169; table 8.1). As such it can offer considerable insights into economic and social life. He has also demonstrated that items such as querns, still often considered to be or little value for chronological purposes, can "be classified on the basis of form, size, use, raw material, and method of manufacture, and that different sizes of querns could be correlated with specific periods.... even when only small fragments were preserved" (Runnels 1994, 165). Runnels goes on to point out how sourcing can provide valuable information about movement, modes of production and exchange.

Dating of chipped stone is usually based upon the methods for working it, or the relative proportions of length, breadth and thickness. The resulting chronologies tend to be in bands of millennia thus, although it may be possible to show that certain types of activity may have taken place in parts of a region, it will be much more difficult to demonstrate the existence of tool kits comprising artefacts which were truly synchronic in their use.

For some lithic artefacts there are well-established functional typologies, based on experimental work in which classes of artefact have been tested for the cutting, scraping, boring etc., of various substances; because of the constraints within which particular shapes of particular stone materials perform and because, in exceptionally rare circumstances, residues of the object material have been found on artefacts, a limited range of wear patterns has been recognized. Even without optical assistance certain types of wear may be visible. At Maddle Farm the vast majority of flints were recorticated, so that fresh breakages due to the modern plough were easily distinguished from old breakages which were more likely to have occurred during their original use, at points exposed to repeated pressure (Gaffney & Tingle 1989, 46); "localised re-sharpening" was another indicator of wear. In general lithic artefacts such as flint, obsidian, volcanic rocks, grits suited to milling etc., are the most durable of all. However, their visibility varies greatly. Where secondary and tertiary flint flakes have been imported they have the advantageous property of being shiny, particularly when rainwashed, but against a background of clay-with-flints or chalk worked material may be hard to distinguish from pieces altered by natural or agricultural processes. The colours of other rocks may easily blend against the background, or they may appear to be recently used building material. Many fieldwalkers seem to require special training to "see", let alone collect or record stone.

4.1ii   Ceramics

There are great variations in the durability and visibility of baked clay artefacts. Firing techniques are the most important determinants of durability, which in turn influences the size of the sherd or whether or not it survives at all. Jeremy Taylor has suggested that ceramic fabrics in the ploughzone reach an equilibrium after which there reduction by breakage and ware is much slower (Carreté et al. 1995, 267; table 9.7), but this is only true of well-fired material. Some early sherds break down when exposed to sun or rain and those which survive to be recovered from the ploughzone may be reduced to bead shapes, in no way reflecting the form of a vessel. The improvements of firing in some Mediterranean areas have made ceramics a viable target for fieldwalking for periods as early as the second millennium BC. In contrast, most north European materials have a low survival rate until the expansion of the Roman influence or conquest. In most of Africa ceramic survival is later still.

For pottery there is not only formal but decorative and material variation over time, facilitated by the plasticity of the medium. Consequently, it is possible to apply tighter chronological control, although the character and resolution of the latter varies greatly from one period to another. In some regions this deficiency has been due in part to the unwillingness of some archaeologists to make the maximum use of coarse wares when fine wares have been present, albeit at a much lower incidence rate. In Britain the prevalence of coarse wares has prompted their routine analysis, and detailed typologies have been elaborated from seemingly unpromising material. Indeed, it has been suggested that fine wares are unreliable means for dating activity since they may have been curated over long periods.

Where it survives well within the ploughzone friable pottery may be a good indicator of date and activity, but it may occur at some distance from the place of its initial use. A high instance of pottery often derives from a midden deposit, whilst thin scatters have been attributed to manuring. Careful statistically based modelling may be able to distinguish the significance of variation in distribution densities.

4.1iii   Architectural material

This is a broad, and often composite category. It includes some lithic material, such as wall and roof stone, flooring, presses, etc., as well as ceramics for tiles. Window glass or lead piping may survive. These latter are rare, and are unlikely to fit easily into a statistical analysis, but stone building materials can occur in such volume, and are of a weight, which make collection impracticable. One solution has been weighing and counting in the field, or even making sketch drawings of their locations.

In some areas buildings survive at ground level, giving a strong indication of their original shape and size. This may be an indication of recent erosion of old deposits, or of abandonment in an area where, and when, the demand for land was not acute. But architectural material is not solely the by-product of shelter. An integral part of the design of the UNESCO Libyan Valleys Survey was the mapping of walls. Their distribution and function was considered to be of great interpretative significance, and the detailed records were used to interrogate the following seven hypotheses:

> "1. They concerned the capture, storage and redistribution of surface water for human and animal consumption and irrigation; 2. They related to the control of fluvial erosion, sediment capture, transportation and deposition; 3. They were built as enclosures to control the movements of animals, whether to protect stock inside pens or protect arable land from animals; 4. They facilitated driving stock or game across the landscape; 5. They delineated areas of different land use; 6. They were by-products of stone clearance, whether to clear cultivable soil for crops or rocky hillsides to improve run-off; 7. They were territorial boundaries, defining parcels of land owned or controlled by different individuals, families, kin-groups, neighbourhoods or larger groupings."

(Barker 1997, 192-193)

To answer these questions, building on earlier analysis by Graeme Barker and Barri Jones, David Gilbertson and Chris Hunt have presented maps of *wadi* walls and settlements close to them, and then subjected the layout of the walls to statistical tests. Relationships in space are only valid targets of analysis if they can be shown to be related in time, too. It was noted that sometimes particular modes of construction were common to the walls and the remains of nearby buildings. Once this observation was placed alongside geomorphological data it was possible to identify earlier and later Romano-Libyan, medieval/post medieval and recent periods of construction. Detailed case studies of two *wadis*, Mansur and Umm el-Kharab, were presented in the report, the latter including a complete larger scale *wadi* system (including tributary *wadis*). Quantitative analysis was used to examine clustering in the distributional and locational characteristics of the walls (Barker 1997, 211-215), and whilst there were patterns the analysis demonstrated that natural factors, such as topography, were by no means alone in determining the layout of systems, leaving room for the introduction of socio-political, as well as economic considerations. By setting the data against the explicit list of questions the authors were able to show that "in addition to the predominant needs of water and sediment control, walls were also constructed to demarcate fields (which we assume were mainly for arable crops), to enclose stock and aid their management, and perhaps to demarcate territorial boundaries" (Barker 1997, 225). The work on the walls formed a key element to Barker's political and economic summary narrative at the end of the first volume of the report, again presented as answers to specific questions.

4.1iv   Environmental data

The variety and quality of environmental data is vast. It has the potential to provide us with information about vegetation in the landscape (natural or indirect consequences of human agency), crop regimes, domestic and wild fauna, diet, animal and human health and methods of agriculture. Appropriate analysis of sufficient samples will shed light on "the adaptation of people to changing environmental conditions, development and implementation of new subsistence practices, influxes of new peoples, or trade in exotic goods" (Hansen 1994, 173). The data include pollen, charcoal and sooting, mollusc shells, bone, material preserved in waterlogged deposits, soil chemistry and grain size, impressions on ceramics and much else besides. But there are inherent problems with many of these data types. Of those which might survive on the surface, dating can only be achieved by expensive specialist means. Some may be collected by techniques such as augering or test pitting, but will again require specialist analysis. For these reasons most surveys have collected only a few samples which cannot possibly be used to make reliable general pronouncements about a region. It is striking that even in an area which has frequently been the object of surveys and excavations a zooarchaeologist has complained that the record is impoverished and biased to certain types of specialist, often religious, site (Reese 1994, 192-193).

Marijke van der Veen has been involved in two important attempts to redress the balance, on the one hand providing pre-arranged specialist input for several otherwise unconnected excavations within a particular area, targeting a particular period; on the other hand working within a regional project and targeting a selection of particular deposits for sampling, so generating a far more reliable regional account than has appeared in most other reports (in Barker 1997).

Geochemical survey, typically by examination of the phosphorous levels and magnetic susceptibility of the soils as indicators of the presence of organic residues, is the technique with the most obvious regional application since augering in the field is a reasonably swift procedure. However, recent experiments with samples from Butser Ancient Farm led to the conclusion that neither of these techniques could be relied upon to be sensitive to enhancement of background chemistry in certain soils, unless specific "biomarkers" were targeted. Only by using combined gas chromatography and mass spectrometry to identify a marker particular to herbivores which was inherently resistant "to microbiological and chemical degradation on structural grounds, and tu leaching due to their hydrophobic nature" (Evershed et al. 1997, 493) were they able to achieve significant variation between manure enhanced and non-enhanced areas of land. The multilayered analysis of deep soils on mainland Orkney looked closely not only at chemical composition, but at the structure of the soil. Ian Simpson (1997, 375) was able to identify the components of manures and the mode of tillage.

The problems inherent in the nature of the data can be overcome, but they require additional investment of field time and specialist expertise. Work over the past two decades has demonstrated that the best results can dramatically enhance regional accounts, sometimes undermining or supporting accepted narratives. All regional research surveyors should at the very least seek the advice of archaeo-environmental scientists. Ideally, specialists should be involved at all stages of a project.

4.2 The meaning of "contemporary"

"The fact that evidence from coarse wares and transport amphorae could not be included in the present study is regrettable, since it deprives us of knowing if the chronological distribution of these categories was congruent with that of the fine wares" (Dietz et al. 1995a, 453)

John Lund's evident embarrassment is entirely appropriate in an otherwise exemplary documentation and analysis of the methods and results of the Africa Proconsularis project. Sadly, although it has become *de rigeur* to treat the surface artefact distributional continuum within a statistical framework, there has been no consistent attempt to treat the temporal dimension in similar fashion. It is not only chronological resolution which is lost, but the opportunity to represent activities in time, associated with the particular artefact classes. This presents problems on a multi-phase, well stratified site which are greatly exacerbated in regional survey precisely because of the types of question survey must attempt to answer.

Foley's solution was to suggest a timespan during which the material he collected had been discarded, on the basis of the artefact-type range. He then proposed a range of models for subsistence derived from ethnographic observations from which it was possible to infer the duration of camps, and the frequency of their establishment (Foley 1981, 109-112). From this he made predictions about the frequency and distribution of surface material for each model. But the social models described by Foley are all static; they allow for no technological or internal structural change. A static pattern may reflect actuality in some cases, but by generalising time and simply linking artefact distribution as averages to that span, the potential for informative contradictions (negative feedback) within the model is greatly reduced.

Schofield accepts this limitation as in the nature of the beast. He has argued that it is precisely with these "patterns of continuity and accumulation that regional survey should be concerned", and goes on to state that "Surface collection cannot be expected to produce the residues of specific moments in time" (Schofield 1991a, 119). He is not, of course, claiming that survey is incapable of showing any chronological variation, merely that it will be at a low level of resolution.

Most survey reports include settlement or cluster/ concentration maps, usually on a period by period basis. No exception should be taken to them if they are merely a means for locating areas of archaeological interest; but it is seriously misleading to use such maps, with the "period/phase"/spatial analysis from which they are generated, within the interpretative structure. If we are to make inferences about population, interactions between settlements, association of land with settlement, political units etc., it must be possible to demonstrate with certainty their contemporaneity, or to illustrate their potential for co-existence in probabilistic terms. This condition was fully understood, though by no means fully met, on North Keos, where "relatively ephemeral economic and social transformations recorded in historical and epigraphic sources" were not easily matched with the archaeological survey data (Cherry et al. 1991, 329). With an ironic honesty the authors recognised that without the check of history interpretation for the prehistoric period was less problematical!

At Stilo much more time was invested in the analysis of finds in an attempt to ensure the reliability of the broad band chronologies they were employing, covering the Early Neolithic (Stentinello) to the Early Bronze Age, and to show sequential change within them. Every prehistoric sherd collected during the survey was coded according to the size and percentage of inclusions, surface treatment (including ware or total erosion) and colours of cores, margins and surfaces. Histograms representing variations in these classes of information showed a marked variation in the proportion of different sized inclusions from one period to another. Thus, in the Early Neolithic there tended to be a greater range of inclusions, particularly the medium fine and medium coarse, but by the Late Neolithic assemblages are dominated by coarse inclusions, a tendency which has become even more pronounced by the Early Bronze Age. The histograms from eight sites were presented in a sequence descending on the page from few to predominantly coarse inclusions, so illustrating a more graduated chronological sequence (Hodder & Malone 1984, 134-135; fig. 8). The authors recognised that other factors, including location, might play a part in the choice of fabrics and tested for this by dividing the survey area into three arbitrary zones, and plotting the sites from each zone against inclusion ratios and sherd colour. It emerged that there was sufficient patterning to suggest that the nature of fabrics was a matter of locality of manufacture, as well as date (Hodder & Malone 1984, 136; fig. 9).

The Tarragona project deliberately eschewed historically determined typology, preferring an archaeological, stratigraphically-based phasing of their ceramics (Carreté et al. 1995, 57, 63), thereby assigning dates to a much greater range of vessels than in most other surveys; but time was still apportioned in multiple centuries. Although it may be possible to demonstrate some of the contemporary links between various places and institutions, the limits of the historical record are illustrated by a demographic survey carried out within the Nemea Valley Project (Sutton in Wright et al. 1990, 596-603. See also Sutton 1994, 319-327). There the fortunes of the valley's inhabitants have been regularly influenced by external events which have provoked sudden expansion and contraction, and changes in preferred location. If, for instance, a "Modern" period had been defined by archaeologists of the 4th millennium AD, comprising the nineteenth and twentieth centuries (and it is not uncommon for reports to produce "maps" of Late Neolithic or Early Bronze Age settlement representing a span of upto a thousand years), the valley would seem to have contained a moderately dense pattern of nucleated and dispersed settlement with a population living by the herding of sheep, cereal-growing, wine production, milling etc. The total valley population might be estimated at between 5-10000, and high and low status areas might be indicated by artefacts relating to transport (carts to cars) etc. In fact, over the period records show that the population had grown from around 700 to 5000, swollen by people from neighbouring areas, drawn by improved roads which not only brought in new ranges of goods, but which provided opportunities for roadside businesses. The consequent sudden relocation of preferred settlement areas would make it difficult for our future archaeologists to discriminate between phases of settlement, so that a chronological dislocation might easily be recognised as variation in status and function.

The excavation of several sites within the valley might eventually identify enough phase sequences to demonstrate the chronological relationships of the various hamlets and villages, so long as the past bias towards "'unusual' and monumental sites" at the expense of "'ordinary' and less visible sites" (Shennan 1985, 62) was avoided; without such information, the view from the surface is likely to be very different. The distribution map would show a prosperous trading elite, whose wealth probably derived from the exploitation of an impoverished agricultural hinterland. Simply, multiphase regional movement has been compressed into one period, so eliminating the chronological distinction which would have led to the inference that the road actually precipitated a general change in settlement distribution.

Can we decompress a "period"? Or are we condemned to work over a *longue durée* within which we cannot represent change, but which is bounded by perceptible technological variation at its beginning and end? It is clear that we exercise greater control over time when it can be represented stratigraphically; even if we cannot measure the rate of change against an absolute scale, we can at least look at the relative distributions of artefactual and environmental data from phase to phase. Within a site it may be possible to show that certain events, preserved in the shape of their physical impact, may have occurred simultaneously, or within minutes, hours or days of each other. However, even within a small British Middle Bronze Age enclosure, within which three buildings have been detected, it may be difficult to determine which were in use at the same time, or if one replaced another. What may be feasible are estimates of the probability of synchronicity on the basis of the proportion of material within fills compared to the proportion of similar material in layers. The ideal stratigraphic reference might be an undisturbed midden.

The excavation of sediments and cultural strata has formed a key part of a prolonged and ambitious first stage of a survey of the Jama Valley, Ecuador. The project coordinators have given a high priority to establishing as detailed a chronology as possible by drawing upon three classes of information: sedimentation (in particular, by seeking to correlate ash layers with volcanic events), radiocarbon dating and stratigraphic and seriational analysis of ceramics (Zeidler & Pearsall 1994, 103). Although surface samples were collected from the focal site of San Isidro (Zeidler & Pearsall 1994, 77) the more significant archaeological work of this stage was the digging of test pits there, and at a number of other previously recorded "sites" in the valley (Zeidler & Pearsall 1994, 79-95 and 99-109) with the explicit intention of establishing a chronology for the whole of it. Clearly volcanic ash deposits are not a generally occurring feature of holocene archaeology, but the use of test pits to obtain material for radiocarbon dating and for establishing a ceramic series is widely applicable, where they are considered to be an apt use of resources. If radiocarbon is to be used as a means for dating typological variation it is surely necessary that the material for which each date is obtained is clearly linked with a substantial sample of the artefact types to reduce the likelihood of inappropriate association. By the same token, several dates ought to be obtained for each stratum within the series.

In the Jama Valley three volcanic episodes punctuate broad-band chronologies, which were refined with radiocarbon dating. Stratigraphic ceramic analysis allowed the subdivision of periods, and this process has been continued to produce a substantial comparative database which will greatly enhance the chronological resolution achieved for surface collection analysis during the second stage of survey.

Zeidler and Pearsall have emphasised the mid-points of the Jama Valley ^{14}C dates. Elsewhere the approach to the technique's application in regional survey has been very different. Nelson, using a substantial body of radiocarbon dates, has shown in bar chart form the probability of the coexistence and sequence of different structural elements of the mesoamerican epiclassic site at La Quemada, Mexico. He has achieved this with a weighted distribution of 75% of the sigma range across dates, using uncalibrated dates (Nelson 1997, 100). With radiocarbon dates from other sites within the region he was able to construct a bar chart representing the growth of satellite development around the central settlement (Figure 4.2; Nelson 1997, 103). Clearly the widespread use of carbon dating is not a practical option for surface-based survey, but it may offer a model for the use of diffuse time as a means of assessing contemporaneity on stochastic principles.

The issues for surface survey are: 1) is there a sufficiently large sample size? 2) Does the range of artefacts include material of sufficient typological distinction? 3) Is it feasible for a probabilistic phased sequence to be generated, either by reference to existing typologies, or by selective excavation?

Sample size will depend on the area of collection units, the periods covered by the survey, and other variables. If a collection unit tends to generate too small an assemblage it may be possible to increase the collection unit area (or by amalgamating collection units during analysis), but only at the expense of spatial resolution, in turn leading to restrictions on the range of questions a survey may be able to address. This dichotomy between spatial and temporal resolution may be mitigated by the use of percentile distribution ratios (Carreté et al. 1995).

The chronological variation of fine wares was well understood in the Sergemes Valley, allowing the division of diagnostic shards into 50 year blocks which were used to suggest the foundation and temporal span of sites in graphical form. A series of small scale maps showed the expansion and contraction of settlement distribution, and were fairly effective in illustrating topographical preferences over time. Unfortunately the failure to create a diagnostic coarse ware series detracts from the value of this exercise (Figure 4.3; Dietz et al. 1995a, 782-785).

The ceramic chronotypological range has often been underused. In Mesoamerica and the North American South West research into typological variation has been a focus for serious research but it is only in the last decade that in Mediterranean surveys, for instance, there has been a trend towards detailed analysis of coarse wares. Still, in the great majority of recent reports, broad categories are based on fine and imported wares. Part of the problem lies with the omission of much excavated ceramic material from published accounts, so that projects must either return to archives or resort to excavation to develop a stratified sequence. On the positive side, it is clear that additional investment of resources in this work can be repaid by greatly increased resolution (Carreté et al. 1995; Zeidler & Pearsall 1994).

Even with more detailed chronologies there remain problems of resolution; but might the diffusion of time employed by Nelson (1997) for ^{14}C dates be applied to artefact attributes? The first problem is the construction of an appropriate chronological scale. If, for instance, we were to have around 20 calibrated dates for material found in sealed contexts with Wessex/Mid-Rhine beakers in Southern Britain which ranged from 1900bc, +/- 150, to 1700 +/-100, and around a hundred findspots and sites in the Wessex area, we might attempt something akin to the following analysis. (1) The span of the two centuries from 1900 to 1700bc may be a product of variation in the ^{14}C analysis, of residual use, or of redeposition. (2) -After plotting a frequency curve it is noted that 75% (a total of 15) of the dates are centred within a 50 year span. (3) Allowing 25 years per generation we might argue that this particular type has a 75% likelihood of currency over just two generations. (4) If we extrapolate from the existing samples we may argue that 75% of all new finds of Wessex/Mid-Rhine have a 50% chance of having been deposited within the span of one particular generation. (5) Allowing for the 25% of the original sample which lay outside the modal range, we might argue that in a map showing the distribution of all findspots of this vessel type there was a 37.5% probability that deposition had occurred within the same generation.

Sebastian Heath (1996) has attempted something similar by representing a probabilistic date range/volume of ceramics

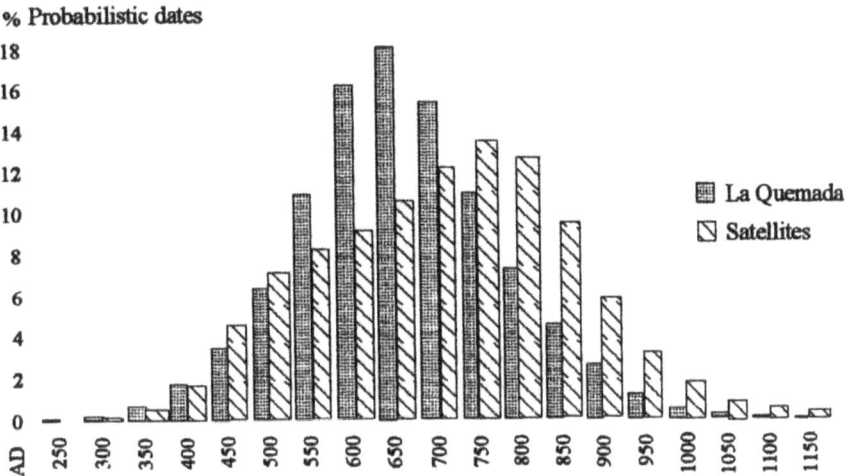

Figure 4.2. Probabilistic dating, La Quemada, Zacatecas. The chart shows the distribution of radiocarbon-determined dates for La Quemada compared with two of its satellite settlements (from Nelson 1997, fig. 8)

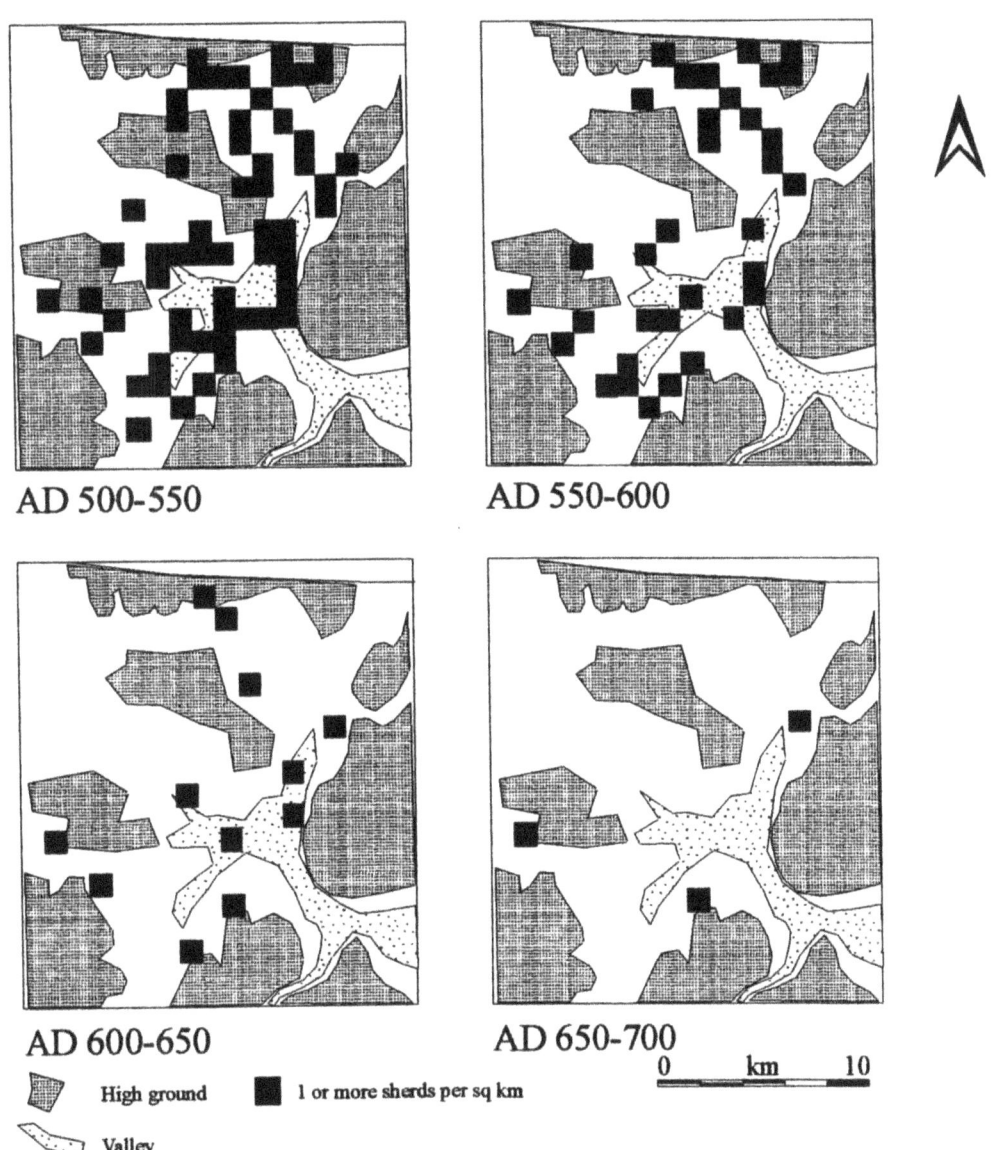

Figure 4.3. Distribution plots in 50 year bands, Africa Proconsularis. Four of fourteen plots representing the incidence of datable wares over a sequence of 50 year spans for each 1km2 of the survey area (from Dietz et al. 1995, 785)

from the Pylos survey displayed across a scale in units of 100 years. Alternatively, if a survey possesses a wide range, and multiple samples, tied closely to specific groups of attributes, absolute time may be an option. More often, much of the span will comprise a time sequence within which the distribution of data is relative, with particular attributes occurring over more than one phase.

In this scheme the relationship of a particular attribute to a specific phase is weakened in favour of the proportional presence of several attributes over a number of phases. For the method to be effective attributes will have to be broken down into a wide range of subtle, but measurable, stylistic inflections in the manner attempted to good effect by Rosamund Cleal on Cranborne Chase (Barrett et al. 1991a, 137-150). Equal attention to detail should be applied during excavation of test pits and, if at all possible, to the selection of deposit types. Undoubtedly the optimum type is an undisturbed midden. As with other layered deposits, what may be recognised as a single stratum may have been developed over many years of consistent formation activity. Therefore it is important to subdivide deep, apparently homogenous, layers by the employment of an arbitrary scale to determine the depth of intra horizonal spits.

Different parts of a large midden are likely to grow at varying rates, and no one part of the midden is likely to contain the full typological sequence, although it should reflect a general order of deposition, within which residuality will feature as anomalous distribution.

Ultimately Chronological resolution will depend upon close stylistic analysis of artefact, and chronological reliability will depend upon the size and comprehensiveness of a sample.

4.3 What do distributional data tell us?

In most cases archaeologists will approach the survey of a region fore-armed with at least some typological information. As has been emphasised in the previous section, the chronological resolution of that information will largely determine the potential for distribution analysis. Without such information a survey can do no more than plot areas that look likely to be of archaeological interest. Before moving on to the role of analysis in narrative it is necessary to deal with pitfalls which may have flawed the data at the point of collection, or which may occur through insufficient statistical rigour.

4.3i   Error and illusion

It is important to distinguish between variations in practice that have occurred through the development of the discipline, or because a project has meagre resources, and those which come about through practice which falls short of contemporary standards. The decision to walk in parallel lines set at 50m apart would have been regarded as perfectly acceptable practice twenty years ago, and would still be regarded by some as a reasonable compromise between extent and intensity of coverage over a substantial region when resources are limited. Less than twenty years ago there were those who considered that an average fieldwalker could scan a 5m wide strip (Bintliff & Snodgrass 1985, 130); reports since that time suggest strongly that about 2m is the absolute maximum which can be expected. At the time of its use the presumption of a 5m wide field of vision was part of the discipline's evolution; if this assumption was made today it would be fair to describe it as bad practice. Unfortunately there is no honest means for adjusting the data to represent the rate of finds as they would have occurred in a 1m or 2m wide strip.

Although the Amboseli Park survey has been hugely influential, it is doubtful whether anyone would feel inclined to gain total representative coverage of a region in Foley's manner, since the extremely small collection of artefacts from very small, widely dispersed units has a profoundly depressive affect on all but the most limited analysis of subsistence strategies. It would be an inappropriate use of resources for this grand and important experiment in survey to be repeated!

The issue of negative evidence is dealt with below (*4.3 ii*) but there are other matters which routinely affect rates of recovery. The most obvious, and utterly unavoidable, are variations both in visibility conditions and in the performances of fieldworkers. There had been some work on these issues in North America but until Shennan took the bull by the horns by "substituting quantitative results for the subjective speculation" (Shennan 1985, 44), most reports were inclined to a sheepish assertion that visibility and performance had little material effect on their results.

Shennan distinguished between "Field" and "Walker "Effects". The former comprised two groups of independent variables, specific to the needs of the survey, listed in Table 4.1.

Table 4.1: Recorded variables

Observation and recording variables (potential distortions)
 1. Soil moisture at the time of the survey
 2. Ground conditions of field
 3. State of crop if field sown
 4. Light conditions at time of survey
 5. Subjective general assessment of visibility conditions
 6. Presence/absence of deep ploughing into subsoil
 7. Landuse at time of tithe survey (c. 1840)
 8. Landuse at time of land utilisation survey (c. 1932)

Topographic/environmental variables
 1. Type of nearest water source
 2. Surface geology of field
 3. Main surface geology within 2 km. radius
 4. Topographic situation of field
 5. Aspect of field
 6. Maximum slope of field
 7. Altitude of field
 8. Land classification of field (Shennan 1985, 35; Tab 4.1)

The second group had direct relevance to the goals of the survey in that it included variables which might have determined settlement and other activity patterns in the past. As such it does not represent an analysis of error; rather, it is a means for stratifying the data. However, error will only become quantifiable when the relationships between all classes of data have been investigated. The first group does include those factors, other than the objects themselves and the fieldwalkers, which might influence directly the rate of artefact collection. The three most frequently collected materials, post-Medieval pottery, chipped stone and burnt flint, were selected as dependent variables against which to measure the impact of the independent variables. Using dummy variable regression Shennan obtained r^2 percentages to assess the extent to which each class of variable had an effect on the perceived distribution of artefacts as the "mean density of finds/ 100m of walked line for each field" after the removal of all zeros and the replacement of the mean figures with their logs (Shennan 1985, 35-36). The analysis is based on a form of equation which will provide an estimated value for the dependent variable derived from the effects of the independent variables (Shennan 1988, 114-134). From this analysis he concluded that 17-18% of the distribution pattern for the collected material could be explained by distortions of observation and recording, while variations in topography and the environment may have accounted for around 39% of the post-Medieval pottery and chipped stone totals, increasing to 57.3% for burnt flint. After analysing separately the effects of environmental and distorting variables, Shennan concluded that the 6.5% to 11.7% overlap of their influence over the dependent variables (the chosen raw materials) was low enough for a consideration of the effect of the former over artefact distribution, without making special allowances for the distortions (Shennan 1985, 36-37), with the exception of Roman-British pottery, the collection rate of which varied considerably with the light conditions, which was thought to have been due to the colour of the pottery (Shennan 1985, 39).

Having shown that the field distortions had little direct effect upon the collection rate of the commonest raw materials, Shennan decided to investigate whether or not they might have influence on "the order of importance of the different environmental variables, and on their degree of importance" (Shennan 1985, 37-38). By comparing the resulting order of effect, firstly from environmental variables only on each of the three dependents, and then of combined environmental and distorting variables, it was found that certainly in general, and usually in the particular, the latter had little influence over the former.

Shennan's arguments are complex. The implication is that further substantive explanations are required before the percentage of variation in each class of dependent variable (each raw material) can be accounted for. It should be noted that the analysis of environmental information implied some human activity might be in response to biological preferences; but a large part of the artefact distribution had to be explained by other means. These could only be addressed once the effect of fluctuations in the performances of individual fieldwalkers had been ascertained. Using similar methods of regression analysis it was found that variation between walkers might account for 3.2% of the chipped stone distribution, and just over 9% for burnt flint and post-Medieval pottery. However, ranking of the walkers illustrated some sharp variation not only between walkers, but in one walker's ability to spot various materials. For instance, walker 4, who was the second most effective performer out of twelve for chipped stone and burnt flint managed only eleventh position when collecting post Medieval pottery (Shennan 1985, 43; table 4.11).

There have been similar findings, presented in even greater detail by Alex Turner (in Aston & Gerrard 1995, 55-70), from the Shapwick project, with strong and differing biases displayed by individual walkers. Although grid walking was applied at Shapwick only data arising from line walking was subjected to regression analysis, since this gave the widest and most balanced coverage of the fields surveyed. After separate analyses of variation between each walker and in the performance of each walker, for each of the four selected dependent variables (Aston & Gerrard 1995, 66; figs. 2.19-2.22), the results were presented both at the line and field level (Aston & Gerrard 1995, 67; figs. 2.23 & 2.24). Turner was brave enough to name the people concerned, and it is interesting to note that whereas the least experienced walkers tended to have difficulty finding pottery and, initially, flint, they tended to out-perform moderately experienced walkers when collecting tile and slate. The most experienced workers usually produced a good all round performance, with one M. Aston producing the best overall results! Using the same method of analysis as Shennan, Turner concluded that inter-walker variation was less than that noted in east Hampshire, ranging from 0.53% to 4.5% (Aston & Gerrard 1995, 67; fig. 2.23). However, it would be very easy to quarrel with his choice of dependent variables: tile, glass, metal and brick. They were chosen because they gave a "large sample size combined with as uniform a distribution as possible" (Aston & Gerrard 1995, 57). This selection fails to tackle precisely the nature of the raw materials problem; most artefacts in the list are "modern", with more fragments likely to be large, and some with a shiny finish. The rank orders, presented in graph and table form, suggest that these items were the ones which most readily caught the eye of the inexperienced walkers (Aston & Gerrard 1995, 67 & 69; figs. 2.25-2.27). The evidence implies that it takes longer to train the eye to see flints and, most particularly, pottery.

At Tarragona a larger collection unit mitigated against this effect. There, Millett concluded that "As our methodology involved aggregating finds made by individual walkers by field, it seems likely that variations in efficiency between walkers will be reduced" (Carreté et al. 1995, 215). Of course, what is reduced are not "variations in efficiency", but variations in perceived or analysable efficiency. Nonetheless, there is likely to be a genuine reduction in the significance of the variation when the larger unit is used. To test this assumption, without the knowledge of the walkers, a record of who collected what was kept for one field in which it was noted that there was a high incidence of mainly Iberian pottery.

The team comprised ten members, five of whom were considered to be experienced. The analysis was based on a simple principle which assumed that the larger the average size of each collected sherd, the more likely it was that the walker was missing some material. Variation was measured in terms of the deviation of each person's average from the overall survey average. The results indicate a slight advantage to the experienced walkers (by a 6 to 5 margin on simple rank ordering) but there were two very significant deviations from the standard. One experienced walker produced a high sherd size, with a low recovery rate, which was explained as probably an area genuinely of low density, while the second and by far the greatest deviation was produced by an inexperienced member of the team who "was also less successful at collecting pottery"! While the assumption that distribution represented in large units will not be significantly effected by this sort of variation further analysis of the one walker's low efficiency makes salutary reading for projects using small collection units. The average size of sherds recovered by walker 10 (Carreté et al. 1995, 215; table 7.1) was 64% greater than the average for all walkers, and slightly more than 30% larger than those of the eighth ranked walker. While it may be contended that a large average sherd size merely indicates the plough's fresh intrusion upon stratified archaeology, there is good evidence from elsewhere that larger size is indeed a function of the fieldwalker's observation skills. At Maddle Farm, where weight rather than size was the criterion of bulk, the relationship of surface to topsoil and stratified finds came under close scrutiny; the top soil was sieved, so was bound to provide a smaller mean sherd size, but the stratified deposits were not. The mean weight of surface find was double that of stratified finds, which was presumed to reflect "the inadequacy of the collection procedure" (Gaffney & Tingle 1989, 129). If the Tarragona fieldwalker really was of such quality, a survey using her/him, analysing by line or grid sample points within a field, would suffer considerable impact on a distribution map, particularly where the overall rate of collection was low. Ultimately, project directors must choose between a "smeared" view of settlement and activity patterns, or one which is highly resolved, but likely to contain significant errors.

Turner was prepared to admit that while inter-walker differences were small (lower than those found in east Hampshire) it was clear that they brought "a significant addition to the variation caused by the effects of environmental and distorting variables" (Aston & Gerrard 1995, 64). However, other surveys have shown willingness to accept that "The impact of the individual observer on the visibility variability is very small" (Dietz et al. 1995, 127), based on calculations conducted elsewhere. It is even more disconcerting to encounter a survey conducted as recently as that at Kavousi, from 1988-1990, where the preliminary report has offered a very detailed description of field procedures and the somewhat dated theory underlying them, which took note of environmental variables, but failed to raise the issue of inter-walker variation (Haggis 1996, 382-389). Some of these issues would be resolved by close initial supervision and training of walkers; but others can best be addressed by careful statistical analysis, where the appropriate data have been recorded.

Further variables considered have included light, surface moisture, compaction etc. Millett noted that lack of variation in rate of artefact recovery at three different periods of the day in Tarragona demonstrated that light had little effect (Carreté et al. 1995, 216); elsewhere the elapse of time through a day has brought changes which have been explained by fatigue, and reduced concentration. In contrast, at Stilo it was noted that when team members were walking into the sun there was a sharp increase in the maximum dimension and weight of the average sherd, indicating that fewer small sherds were being collected (Hodder & Malone 1984, 130). The likely causes of this discrepancy were the bleaching out of surface relief, and glare. This suggests that in general, in the northern hemisphere, it would be wise to face north in the winter and south in the summer, whatever the collection technique may be.

Once the effects of distorting variables and those associated with fieldwalker inconsistency have been evaluated, environmental/topographic data have an ambivalent status; on the one hand the aspect of a field may contribute to the problem of artefact visibility, but it may also have informed decisions about landuse in the distant past. Similarly, a hilltop may be bare of finds because erosion has moved them downslope but, equally, there may be phases when settlement or use of such topography was avoided. In some respects this is not altogether a problem; regression equations are used to seek the most economical set of relationships between different independent variables, and the degree to which they control a dependent variable. Although we may have trouble differentiating causes within a determined variable, nonetheless we are provided with a means for quantifying what remains to be explained. We have moved from physical data, some of which have been qualitative and have been coded subjectively (light conditions, ground moisture; but note that even these conditions are often quantifiable with current technology - although issues such as angle of light combined with the reflectiveness of the surface may-continue to be a problem), and others of which are susceptible to measurement (size/weight of artefact; angle of slope; aspect etc.), to modelling the mediation of that data via a field collector (the walker), through to a stage at which the remaining explanations seemingly lie in the less tangible areas of interpretation. In this process we have treated artefactual data at a fairly general level, both in terms of object type and chronology, but we have added meaning to these generalised objects with respect to their location at the time of collection, the possibility of that location having been effected by movement in the time between discard and collection and, in some cases, by assessing the effect of the processes of mediation (the relative efficiency of the mediator). In essence we are considering the class A stage of narrative, with some regard for class B.

Error is not restricted to the fieldwalker. The preconceptions of those within the management structure of a project can cause significant distortions of interpretation. There is still a widespread assumption that a high density of artefacts is indicative of settlement; an increasing amount of ethnographic work has shown that this is an unsafe assumption since in many cultures dwelling and communal areas are kept clean and waste is removed to some distance from the camp or settlement (Schofield 1991a, 117; Hodder 1982). The importance of being able to distinguish between a large midden and a settlement may depend on the degree of resolution required by the survey. If the aim is to locate or recognise the internal functions of a settlement failure to distinguish between the two will prove a major flaw. But from the off-site perspective a generalised view of human activity and its location may suffice.

Equally, erosion will cause varying, often apparently contrary effects. Sheetwash will rarely cause a significant lateral displacement of artefacts, but it will remove fine soil particles, so creating the illusion of a high density of finds upslope, while gradually generating a colluvial seal of material discarded close to valley bottoms (Allen 1991, 45). It is disconcerting to note that the Tarragona survey found the greatest number of "sites" on "Knoll, ridge-top or hill-top" and at "Break of slope", and the lowest number on the "Valley floor" (Carreté et al. 1995, 245). Over the past three decades research in southern Britain has uncovered previously unsuspected evidence of Early Bronze Age settlement on valley bottoms; indeed, there is now more artefactual evidence of the Beaker phases from valley bottoms than from elsewhere (Allen 1991, 53). Ploughing will tend to move material downslope, and experiments with resin sherds "tracked over six years in a chalkland hillwash in a valley bottom, cultivated by ard" (Yorston et al. 1990, 69-70) suggested that artefacts might have moved by as much as 5m in one year even during prehistoric periods of cultivation. Conditions of extreme rainfall with a high level of surface flow have moved upto 87% of artefacts downslope in semi-arid, experimental conditions (Wainwright 1992, 233-234). However, whilst it is generally accepted that the raw material and shape of an artefact will determine its relative mobility, there is much less agreement as too which artefacts are most susceptible to erosion. At Paoura, North Keos, it was felt that pottery was unlikely to move as far as obsidian, and that therefore it was a better indicator of settlement (Cherry et al. 1991, 213), whilst experiments at Ashcombe Bottom show clearly that on the chalk downs of Sussex pottery tends to move downslope more quickly (Allen 1991, 47; fig. 5.4). It may be that durability of material rather than mobility was the factor determining distance of movement. For this reason it may indeed be that pottery is a better indicator of settlement, providing that it was discarded close to its place of use.

Prior to the use of more sophisticated statistical analyses, distorting factors in the field were mentioned and, in the case of vegetation cover, their extent was usually estimated in terms of the percentage of ground covered. The interest in the mediator of physical data is of more recent origin, usually having gone no further than references to "experienced", "post graduate", "student" and "amateur" fieldwalkers. Although the surveys which have conducted such analyses have discerned little significant effect we must remember that the variables we deal with, and the relationships between them, are very prone to fluctuation (volume/density of finds; quality of labour; ground/light conditions); it is only through repeated studies, and with new means for assessing them, that researchers will be able to present data and subsequent narratives with confidence.

4.3ii The meaning of nothing

At first sight this heading appears to indicate a concept which is the opposite of thing, of entity, but in terms of survey logic it may be viewed as a two-dimensional spatial concept. This "nothing" or "low incidence" is also part of the range of arbitrary data categories which describe the target population of a survey. *Nothing* can be paradoxical; due to the eroding nature of the Amboseli Park surface Foley felt secure in suggesting that much of the areas where artefact density was low reflected a general lowering of the frequency of human activity, compared with other portions of the surveyed area, over some 2000 years. In contrast, confronted with an area of low density within a probable Roman estate, where the soils were optimal for arable agriculture, Carreté et al. came to the conclusion that the *lack* of finds were an indicator of a particular human activity. The land was too far away from dwellings to be covered with manure incorporating domestic refuse, but it was of sufficient quality to require little extra nutrition for arable use; that little might have been provided by the seasonal grazing of animals. Evidence from elsewhere in the survey suggested that even marginal land was being used during this period and it was reasoned that it was very unlikely that the best arable land in the territory would not have been utilised (Carreté et al. 1995, 272).

Other forms of negative evidence can be due to post-depositional factors. All the artefacts (and features) on an area of raised land may be removed or destroyed by intensive ploughing, whereas alluvium or marine silts may form a comprehensive mask over archaeological residues which could only be removed at prohibitive expense (the work at Seamer Carr is a rare and successful attempt to locate such residues; see *3.3xb* above).

Negative evidence can be chronologically biased. There are periods during which we know there must have been a human population, but it remains hard to detect because the way of life included few durable tools or other artefacts. This problem is notorious amongst British archaeologists with an interest in the Early Medieval period. Only by introducing forms of geomorphological survey, such as test pitting and augering, will we be able to evaluate the nature of negative evidence.

4.3iia Soil and distribution

Soil has a dual function in the narrative sequence. At one level a particular soil which has been formed through direct, or indirect, human agency is an artefact, the components of which may best be treated as attributes no less susceptible to classification and statistical analyses than the decoration of pottery. Equally, soil is the medium within which other artefacts - the fragments of vessels, tools and other waste - have been stored in space or redeposited. In a stimulating discussion Armando de Guio has rightly asserted that sometimes it is possible to infer the probable stratigraphic origins of artefacts from the components of surface soils (Christie 1995, 18-19).

Recent work of the South Cadbury Environs Project has explored this association with intensive work in an 18ha field at Sigwells, Charlton Horethorne. Fieldwalking revealed that Romano-British surface finds tended to concentrate on or close to areas of soil which appeared grey or dark brown after ploughing. This observation is not new to the literature, but it presented an opportunity to explore the strength of the relationship. The previous fieldwalking had been in 20m grids, but it was decided that for the purposes of experiment two further techniques should be used. A programme of shovel testing by sieving 30 litres of ploughsoil at every 20m along lines at 20m intervals was set against the results of 19 $1m^2$ test pits (one pit per hectare) which were dug down to natural. Soil characteristics (colour, texture etc.) were recorded according to a given set of numerical values in all shovel tests, so as to limit subjective variation.

Analysis has not been completed, but a preliminary review suggests that whilst it may be unwise to treat as absolute the correspondence of soil type with frequency of finds, there is a strong correlation. Of the nineteen test pits only three contained dark brown plough soil. Two of these ploughsoils produced Romano-British pottery, but in low quantities belying the volume of material found under them, whilst the other had none (Figures 4.4 and 4.5). Only 50% of the other sixteen (medium brown) ploughsoils produced Romano-British pottery, and in a proportionally lower

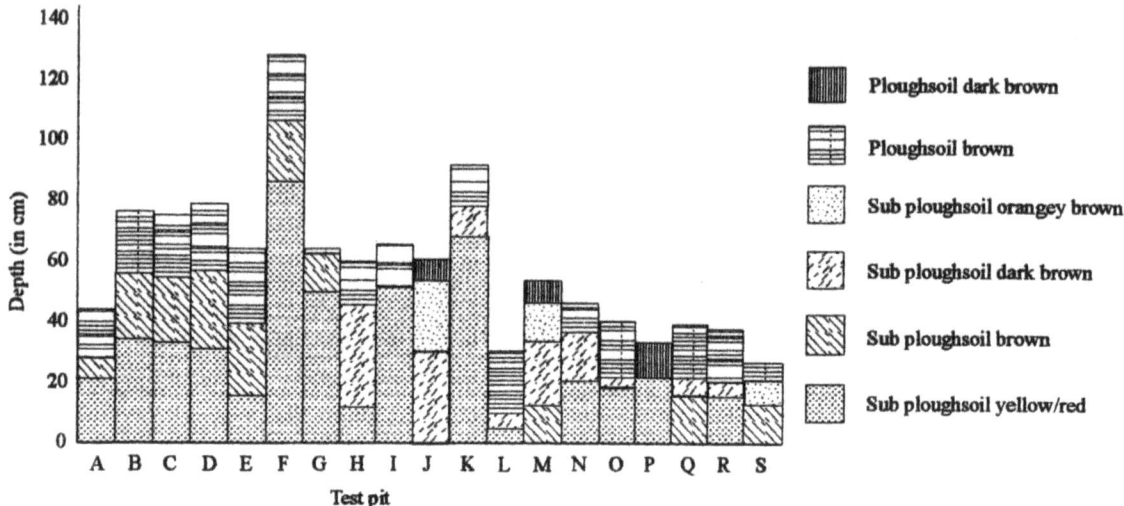

Figure 4.4. Depth of each soil type per test pit, Sigwells, Somerset

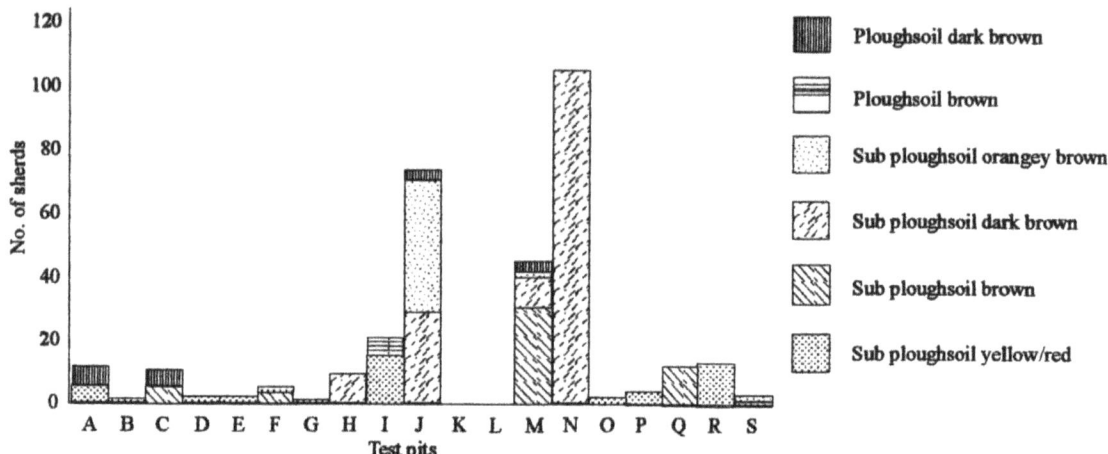

Figure 4.5. Distribution of Romano-British pottery by soil per test pits, Sigwells, Somerset

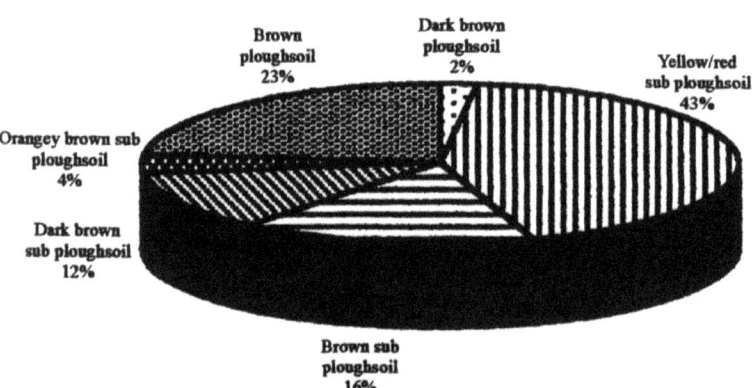

Figure 4.6. Relative proportions of soil types in all test pits, Sigwells, Somerset

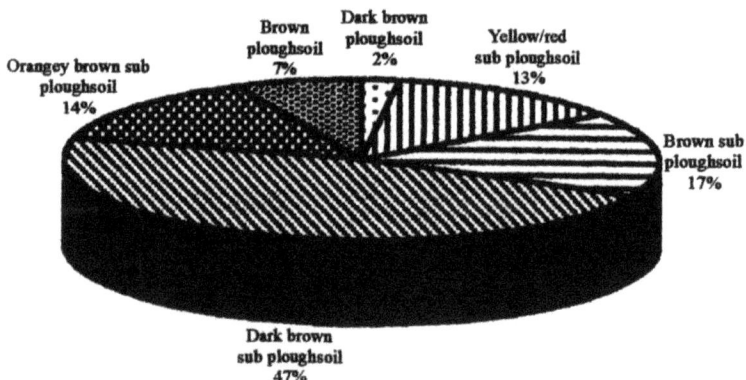

Figure 4.7. Summary percentage of Romano-British sherds per soils, Sigwells, Somerset

volume (Figure 4.6). The nature of the correlation between dark brown, sub-plough level soils and Romano-British pottery was more surprising; only four out of nine such strata produced any, although those which did so produced 55% of the total assemblage, despite amounting to only 15% of the soil by depth (Figure 4.7).

Two further inferences which may drawn from the statistics concern the incidence rather than the volume of Romano-British pottery by soil type per test pit. Only the Sub-plough dark brown soil had a lower than 50 rate of incidence, at 44%. Three of the five types had a 60% or more association. Within the soil types the rate of recovery of finds per cubic metre varied considerably. This pattern suggests that the size of test pit was capable of reflecting subtle changes in distribution. It also suggests that the yellow or red soils were sometimes open, probably cultivated, during the Roman period, although it is possible that there has been downward migration of finds due to faunal activity.

De Guio's proposition is interesting, but it requires more widespread testing before archaeologists surveying the surface presume a strong association between identifiable soil types and finds.

4.3iii   Quantity and quality

It is probable that most surveys now adopt the iconic analytical approach proposed by Clarke, and several have made tentative steps towards the symbolic by offering hypotheses which may be susceptible to testing in the field. Unfortunately, as Baxter has noted (1994, 224-225), it remains rare for reports to present the full line of reasoning. Exceptions include Foley (1981), Shennan (1985), Gaffney and Tingle (1989), Carreté et al. (1995) and Steinberg (1996); another example is in Aston and Gerrard (1995), but this is an interim report of a project in progress. Most recent surveys do explain their criteria for dots on the map which represent "findspots", "small sites", "large sites", etc. However, these are usually extremely arbitrary definitions of what are Abnormal Densities Above Background Scatter (ADABS); a surface concentration might be constituted by some 20 sherds in an area of 1 hectare, most of which are undifferentiated coarse wares. Three of these might be diagnostic finewares of the same phase as each other. It has been decided that 15 sherds per ha represent a medium-sized site and over 30 sherds a large one, so this becomes a medium-sized site of phase X, and duly earns its place on a distribution map as such (for remarks on this problem see Mattingly 1992, 99; for examples of it see Chapman et al. 1996, 50; Barker 1995, 138, 162, 184; etc.). The worst examples fail to state the number of sherds as a ratio with respect to a given area of land, and make no account of durability and variable rates of supply of pottery from different phases.

Of those good exceptions listed, only one aspired to class G narrative, and that was assisted by being based around the major Roman provincial capital of Tarraco. Much of the reasoning behind the statistical analyses employed for choosing between models for patterns of encampment and hunting activity in the Amboseli Park has been set out above (2.2i) and need not be repeated here. However, the following section will attempt to follow statistical analysis of field data which has been used to generate narratives.

4.3iiia   East Hampshire, Britain

This programme of fieldwalking in regularly spaced transects, conducted in 1977 and 1978, was one of the first in Britain to follow an offsite model of collection and analysis (Shennan 1985, 8-9). The title of the report, "Experiments in the Collection and Analysis of Archaeological Survey Data" at once made explicit an aim to place survey within a more rigorous framework, while implying that the narrative produced could not represent a definitive account of East Hampshire's archaeology. Perhaps its greatest importance lay in the advancing of Foley's use of statistical analyses, by including the effect of the fieldwalker on collection (*4.3i*, above), and by breaking down artefacts into more meaningful categories, so generating a more detailed narrative.

The analyses are divided into a series of discrete, period specific, units based on the characteristic artefacts of each period. Each analysis is an attempt (Shennan 1985, 53) to account for the deviation from normal distribution which was not accommodated by environmental/topographic, distorting and walker variables. "Chipped stone" was the only widely occurring raw material which could be used to investigate activity from the Mesolithic to the Early Bronze Age. Standard observations concerning length:breadth ratios suggested that the total collection contained surprisingly little of the former period, bearing in mind that work in neighbouring areas had suggested Mesolithic activity over most of dry-land Hampshire during the period (Shennan 1985, 54).

A crude discrimination of function across the landscape for this whole long period was attempted by investigating the relative distributions of touched and retouched flints. It was assumed that places with an abnormally high proportion of retouched flakes were "those most likely to represent places where maintenance activities were carried out and the tools ultimately discarded, as a result of occupation either over a long period of time or recurrently: settlements in other words" (Shennan 1985, 58). To gain greater chronological resolution a sample of the total flint collection was subjected to detailed analysis. All the flints (a total of 2682) recovered from 55 fields had the following details recorded: The field number; Easting and Northing grid references of field; line number within field; type of piece (flake/core/rejuvenation flake/Shatter); length and breadth of piece; platform angle of piece; extent of cortex (primary flake/secondary flake/tertiary flake); hinge fracture, whether present or absent; presence of use (none/use/retouch); retouch type (none/edge/unifacial/bifacial); patination, whether present or absent; colour of raw material (Shennan 1985, 59; from table 5.2)

While it proved relatively easy to distinguish Mesolithic material, there was, in general, no strong clustering of greater length to breadth ratios which received opinion takes to indicate Early Neolithic against the proportionately broader Late Neolithic/Bronze Age industry. The "widespread but thin distribution" of chipped stone, in a landscape lacking "strong ecological constraints on the location of sites of various kinds" fits a model considered by Foley (1981) "in which the archaeological evidence of mobile occupation accumulates over time" (Shennan 1985, 70).

Although the number of finds collected from individual fields provides a much larger sample size for each collection unit than that for Amboseli Park, and although there was a greater number and range of distinguishable types, the totals of diagnostic artefacts were still too small for functional analysis other than the most general kind; in effect this is a generalised class D narrative. Despite the lack of specific information it is possible to use this information as the basis for a class E narrative, concerning economy or subsistence: that occupation of the region was long term and mobile.

As has already been noted (see above, 3.3ixa) diagnostic surface finds from the 1st millennium BC are notoriously rare in Britain and East Hampshire, with a mean sherd density of 0.034 sherds per hectare, was no exception to this. With so small a sample Shennan was unable to apply regression analyses, but because of the tightly controlled field methodology it was possible to adapt the scale of distribution mapping to enable comparison with other periods. The Early Prehistoric, Romano-British and Medieval periods were compared with the Late Prehistoric by dividing each of the twelve, 0.5km wide by 5km long, transects into 1km lengths (i.e., blocks of 0.5sq km in area) and noting the occurrence of any evidence for each-period within each block (Shennan 1985, 76 and 111; table 9.3). Whilst no claims were made for representations of density or size of settlement, the sample of 60 sections was sufficient for the application of a chi-square test to explore the similarity of distributions between phases. This suggested that the pattern of each succeeding period was most similar to that of the preceding period (Shennan 1985, 111; table 9.4). The tendency for Late Prehistoric finds to occur on geology where the soils, at that time, would have been more conducive to arable agriculture, was explained as part of a pattern of progressively increasing cereal production, starting in the Middle Bronze Age, with more static occupation and a gradual movement from "the areas of extensive clay-with-flints and from Gault clay, for reasons of workability, and from the lower greensand because of the already poor quality of the soil" (Shennan 1985, 77). Once again we are offered a very general class E narrative, but very much on the basis of analogous narratives from elsewhere, from a sample which is too small to allow inter-regional comparisons at a class D level. In terms of the evidence from the survey neither class D nor E narratives are reliable for the prehistoric periods.

Although the Romano-British period generated a larger sample, with a substantially larger mean of 0.234, of the 208 fields walked which met the conditions for statistical analysis only 26 produced sherds of the period. To effect regression analysis where there was so high a zero count it was necessary to reduce the number of independent variables to only five. Of the 53% which they accounted for, nearly 34% was due to light conditions, a marked contrast to the findings at Tarragona (see above, 4.3i). Positive environmental factors influencing the density of pottery appeared to be a southern aspect and the class of land.

There is a general geological, and hence implicitly environmental, division along the north-south central axis. The heavier soils to the west would have been more difficult to work prior to mechanisation, and there is a markedly higher density of artefacts on the east side (Shennan 1985, 82; fig. 7.2). There is a continuous decline in density of sherds from the north boundary, just beyond which lies a significant Romano-British settlement, for around 10kms southwards which continues to the west, but which is sharply reversed on the east. This increase may be due, it is suggested, to the influence of a second settlement area, although none is presently known. Otherwise the east and west trends appear to represent a falling off of the sherd density towards the periphery of the catchment area of the northern settlement, influenced also by the presence of the clay-with-flints (Shennan 1985, 82-83).

Shennan has presented two very general class E accounts for the two prehistoric periods, but for the Romano-British period he attempts a little more detail by looking for signs of variation in economic activity. Taking an arbitrary figure of five sherds and below to represent low density, and above five for high density within a non-uniform areal unit, the field, he noted that there is a 1:2.25 relationship of high, to low in the west and 1:5.8 ratio in the east. After suggesting that low densities are likely to represent manure zones he observed that the number of larger concentrations in the west and east were approximately balanced at a ratio of 4:5. From this he infers that the variations in discard distribution may not reflect variation in settlement density, so much as variation in agricultural practice (Shennan 1985, 83-85).

This analysis is based on a total Romano-British assemblage; Shennan is quick to point out that his argument fails to take into account the full chronological resolution available to him, but any inferences using that information are probably unsafe due to the corresponding reductions in sample sizes, although it may be reasonable to press the case for increased intensity of landuse during the period.

Despite the quality of the statistical analysis there remain doubts about the validity of even the modest narrative claims of the text. Perhaps the most persuasive hypothesis is that concerning the mobility of occupation in the Neolithic and Early Bronze Age; yet even in this case, where Shennan has used conventional artefact proportions to distinguish two possible periods (Early Neolithic and Late Neolithic/ Early Bronze Age) the duration of each chronological band dilutes his statement. Does "mobile" mean a change of occupation site every few months; does it imply transhumance between temporary structures; or is it a matter of generations? He has used Foley's model without supporting his argument with the ethnographic observations and detailed ecological reconstruction which made the gradually changing, and very general, picture of life in the Amboseli Park much more persuasive. While it is appropriate to argue that such a model cannot simply be transplanted it does suggest that statistical applications without something more substantial to test render rather sterile results.

In Owens Valley (Bettinger 1977) detailed ecological strata enhanced a picture where chronology was of a low resolution, but where artefact determined behaviour was quite well focused. In temperate zones the significance of the ecological niche may well have been less cogent in determining economic behaviour, but the consequent loss of definition can be compensated for by avoiding too austere an approach to field methodology. Although Shennan's statistical vehicle may have been state of the art for archaeology, his use of the artefacts themselves was very limited. When the statistical analysis wore thin it was bolstered by borrowing a model of mobile occupancy (from Foley), by borrowing a scheme for discriminating between artefacts (Richards and Bradley), and by confirming observations made from just beyond the peripheries of the survey area (Millett). The geological/topographical strata were not a sufficiently rigorous set of variables to test these schemes, and in place of a site location survey we are left with a findspot survey which has been analysed beyond its means. For instance, the gradual shift from the presumed wooded heavier soils of the west towards the more tractable soils of the east is statistically perceptible, but without competing models the hypothesis that it reflects a particular change in the modes of subsistence is not truly tested.

This method of analysis requires greater resolution for at least some of the variables. There should be no doubting the legitimacy of conducting experiments in collection and analysis, but the field methods selected for East Hampshire were inappropriate; if the objective was to squeeze more narrative from sparse data it failed. Indeed, there are cases where a much less zealous approach to the data can reveal more. A regression analysis of the relationship between surface and sub-surface artefact densities at Knighton Bushes produced a relatively poor result, which was convincingly undermined by the authors with graphs using raw numerical data and description of what had occurred in the field (Gaffney & Tingle 1989, 24). For Shennan's survey to generate a convincing narrative a multi-layered field methodology was needed which would have allowed for the building of a chronologically highly resolved database and for the investigation of the relationship between the surface, topsoil and stratified deposits.

4.3iiib Maddle Farm, Berkshire

The Maddle Farm survey is known best for its analysis of the Romano-British period, although it is useful to compare the analysis of flint distribution with that set out by Stephen Shennan.

It was fortunate, though perhaps a little daunting, that "the entire survey area was covered by a carpet of lithic debitage reflecting at least two millennia of flint working", with a mean flake density more than four times greater than that encountered in East Hampshire (Gaffney & Tingle 1989, 33 & 43). Thus, for broad band chronology there would be none of the problems relating to sample size which Shennan encountered and, if a sufficient degree of discrimination could be achieved, chronological sub-groups might also be susceptible to analysis.

Given that the complete lithic assemblage exceeded 37,00 pieces it was not practical to treat the whole assemblage in the same detail. Once the survey-wide density of flakes had been represented as a number per hectare (Gaffney & Tingle 1989, 34; fig. 5.2), material from certain areas was selected for analysis, from which it might be "possible to determine generalised activity trends on non-site locations" (Gaffney & Tingle 1989, 35). A general picture of the relationship between weights of different flake types (primary, secondary and tertiary) and their distributions across varied geology and topography was represented in a graph/diagram (Gaffney & Tingle 1989, 32; fig. 32) over a 10km, 0.5km wide, transect. Weight was adjudged to be an appropriate variable (as opposed to length:breadth ratios, for instance) since it allowed swifter processing. The pattern suggested that in the west zone of the study area, where the lie of the land is flatter, the variation in weight between tertiary and secondary and primary flakes was less pronounced than in the central and east zone, where the varied topography and weathering had produced exposures of flint nodules.

In six areas, in each of which between 11 and 60ha had been walked, additional features were also analysed: the total and mean number and weight of cores per hectare, and the total number of retouched tools from each collection assigned to the categories scraper, retouched flake, borer, knife, notched flake, arrowhead, fabricator, serrated flake (Gaffney & Tingle 1989, 35-44). These areas were then contrasted with four areas of greater density, designated as "on-site assemblages", using the same criteria and methods of analysis (Gaffney & Tingle 1989, 59-68).

The wide-spaced walking had provided good evidence for significant concentrations in some localities of the survey, and graphical representation of test pitting results showed that the surface data correlated significantly with the ploughsoil. However, it was a goal of the project to investigate not only the relationship of the sampled surface (estimated at between 4-10% at 25m for this survey) to the subsurface, but also the degree to which such a surface sample might represent the whole surface of a given area. Two areas associated with Romano-British activity, Knighton Bushes and Maddle Farm Roman Villa, were selected for an intensive survey carried out in 5m grids, two of which produced flint concentrations, the latter accompanied by 60 sherds of prehistoric pottery. Although no particular relationship or continuity between periods could be shown, beyond the coincidence of location, both demonstrated that intensive fieldwork was likely to uncover densities which might reasonably be described as considerably greater than off-site levels (Gaffney & Tingle 1989, 56-59). A third area at Post Down Farm was treated intensively and then excavated.

The survey was able to discriminate between possible "quarry" sites - areas of flint exposure where the lithic collections had distinct characteristics - and other activity loci, the functions of which could not be ascertained. Within a broad chronological framework discard patterns of diagnostic material suggested that the Middle Chalk valley bottoms were favoured during the Early Neolithic (a view reinforced by a similar siting of long barrows), but that by the Late Neolithic and Early Bronze Age there was considerable activity on the upper hillsides, some or much of which was linked to the extraction of flint, as well as throughout the west zone of the study area. This distribution appears to contrast with the zone east of a barrow group; here, no diagnostic lithics of that period have been found on the Upper Chalk and Clay-with-Flints.

The levels of narrative are still classes E and F, and they remain very general. Nonetheless, some small refinements at the class D level have allowed the placing of some activity. Despite the variety of levels of data (derived from wide-spaced and gridded fieldwalking, test pitting and excavation) the overall structure allowed comparison, each level informing the others with vertically or horizontally derived information. The mode of analysis was lucid and to the point, so that it is very easy for the reader to understand the structure of the arguments. There is nothing contentious in the conclusion, partly due to the limited claims, but partly because of the clarity of argument.

The most ambitious part of the project was the attempt at a reconstruction of the agricultural economy of a villa estate. In effect, the authors attempted to do for an example of the Romano-British agricultural system what Foley had done for a hunter-gatherer economy in the Amboseli Park. Where Foley drew on modern zoological and botanical research, coupled with palynological data, to estimate the biomass of human prey in his target region, Gaffney and Tingle used existing literature concerning the productive capacity of land and domestic animals, including previous estimates for reconstructions of the Romano-British and medieval periods. Their line of argument is clear throughout and deserves to be presented in some detail as an example of how to draw on data which, on their own, have little meaning. The argument stands on an intelligent view of the significance of different artefact distribution densities; the size of the areas of land involved; and their location with respect to particular types of contemporary settlement. At each stage historico-ethnography and experimental archaeology provide pertinent data with which to fill the empty landscape.

By distinguishing between various densities of artefact scatter, it was possible to lend them meaning by association with particular types of activity. The interpretation springs from a set of three assumptions of this kind: "(a) The estate contained a gross manured area of c.211 hectares". The assumption is that domestic waste, of which only the broken sherds remain, was incorporated into dung derived from a dairy herd. A large area was covered by significant, but sub-settlement level, scatters of pottery, (b) "An area of c.39 hectares was available for summer grazing by a dairy herd". Fieldwalking showed that there was an area adjacent to the Knighton Bushes settlement which was "devoid of off-site pottery", but surrounded by the area of manure scatter. Grazing fits well with this negative evidence, (c) "The system carried out at the farm was intended to maximise production". Peripheral foci may "represent dependent cottages or separately sited manure 1 leaps within areas of arable cultivation... a system geared towards efficient land use" (Gaffney & Tingle 1989, 225-227; fig. 14.1).

Using figures from Keith Branigan's work at Gatcombe it was estimated that a minimum of 32 cattle would have been kept on the 39ha of grazing in the summer, but that they would have required byring for between 120-168 days over winter. Several sources were referred to for providing estimated winter fodder requirements, and the volume of manure generated by a herd of 32, as well as the amount of land that manure might cover (Gaffney & Tingle 1989; 227-231)[1]. Unsurprisingly, estimates vary widely. For instance, fodder requirements for each animal range from 0.7 to 1.8 tonnes for 120 days, whilst estimated tonnes of fodder production per hectare go from 3.77 to 4.73 tonnes. Consequently, the estimated total land area required to winter the herd lay between 5 and 15ha. If the wintering period reached the supposed maximum of 168 days the estimate varied from a 7 to 21ha. Using Trow-Smith's estimates for per capita milk yield in Medieval England, when cattle are assumed to have been poorly nourished by comparison with modern domestic bovines, each animal would be capable of producing between 540 and 647 litres per annum. Using estimates of Developing World consumption, compared with Slicher van Bath's assessment of calorie intake by 18th century agricultural labourers, it is assumed that the diet of an average adult on the estate would be around 2,200 calories per day. Without further allowance for variety of diet "the milk carrying capacity of the Maddle Farm herd was enough to feed 12-15 individuals per year" (Gaffney & Tingle 1989, 228; tables 14.1, 14.2).

The argument continues through production of dung from this size of herd while byred, to the amount of land which could have been manured by it (Gaffney & Tingle 1989, 229) for arable production, on the basis of presumed needs of the chalk-soil for nutrients. Drawing on the same literature the figures for dung productivity vary from 5.8 to 8.7 tonnes over the winter, with estimates of the amount applied to each arable hectare varying from 10-20 tonnes. Thus the estimated total area manured ranges from 9-28ha (Gaffney & Tingle 1989, 229; table 14.3).

Experimental work at Butser Ancient Farm has greatly increased the theoretical yield capacity of cereals in the Prehistoric and Roman periods and, consequently, our estimates of the human population which can be carried on an area of land. However, yield will be constrained by the availability of nutrients from the soil which by the Romano-British period would already require intensive manuring, in turn constrained by the volume of manure produced by livestock. Not all the crop will be available for consumption, and the figures allow that at least a third of it was "retained for seed purposes". Gaffney and Tingle considered that the level of manuring required to achieve the experimental yields was too high. Although, Branigan's use of Medieval data may be inappropriate, since it underestimates the stimulation to the economy caused by the need to support a Roman army, as well as the developing trade within and around growing urban settlements in Britain and Europe (Gaffney & Tingle 1989, 240-241), the yield values selected for analysis are within the lower range, at 0.75 to 1.3 tonnes per hectare. The total yield from 10 to 20ha under cultivation would lie between a minimum 7.4 to a maximum of 24 tonnes. If 1 kilo of wheat provides 3,400kcals, and a human adult requires 833,660 per annum, then the estate might be able carry between 30 and 98 people (Gaffney & Tingle 1989, 229-230; tables 14.4 and 14.5).

The other important constraints on yield are manpower and available time. There are only so many days between the time when a crop is ripe enough to harvest and when the grain begins to fall or be spoiled. The literature shows estimates of 9 to 14 man days being required to harvest 1 hectare. If a long harvest period of twenty days is allowed, the varying estimates of land under production allow for a work force ranging from 7 to 20. If the harvest period is shortened to ten days the workforce range must increase from 13 to 39 (Gaffney & Tingle 1989, 231, tables 14.6). The authors doubt that their would have been sufficient flexibility

[[1] **For estimates of winter fodder requirements**: Branigan, K. 1977. Gatcombe: The Excavation and Study of a Romano-British Villa Estate, 1967-68. BAR 44. Trow-Smith, R. 1957. A History of British Livestock Husbandry to 1700. RKP. Balfour, EB. 1975. The Living Soil and the Haughley Experiment. Faber. **For estimates of the area of manured land**: Branigan, 1977. Slicher van Bath, BH. 1963. The Agrarian History of Western Europe AD 50 — 1850. Arnold. Reynolds, PJ. 1987. Butser Ancient Farm Yearbook: 1986. Ancient Farm Project Trust. **For estimates of milk yields**: Trow-Smith, 1957. Slicher van Bath, 1963. Pyke, M. 1970. Food Science and Technology. Weidenfeld. **Estimates of manure production/land fertilised from classical, historical and experimental sources**: Applebaum, S. 1975. Some Observations on the economy of the Roman Villa at Bignor, Sussex. Britannia VI, 118-132. White, KD. 1970. Roman Farming. Thames and Hudson. Branigan, 1977. Slicher van Bath, 1963. Reynolds, 1987.]

within the economic system to allow seasonal migrant labour to be involved, so it is assumed that "only the elderly, the ill and the very young would have been excused work" (Gaffney & Tingle 1989, 232) at harvest time.

The figures used thus far, including areas for settlement, arable, winter fodder and summer pasture, consider between 7.5 and 11% (64 to 94ha) of the estate's presumed extent, and take no account of other forms of economic activity (Gaffney & Tingle 1989, 232-233; table 14.7). Assuming that there was sufficient demand to stimulate productive maximisation, a new series of calculations introduced dung from sheep and humans, although these also brought new constraints, since sheep, along with cattle, would have required supplementary fodder over the winter, which in turn would have had significant implications for the amount of manpower needed. Allowing for this it is suggested that the maximum area under cultivation would not exceed 47ha.

With careful argument it was possible to suggest minimum and maximum populations for the villa and the amount of surplus produce. It was suggested that the population was distributed across three settlement types: the villa, the "village", and perhaps peripheral cottages, which may have had a specialist role within the overall economic structure. The villa complex, comprising buildings of the 3rd and 4th century, was constructed within a pre-existing economic scheme which had included the other settlement components which the ceramic evidence shows were operating soon after the beginning of Roman occupation, almost certainly in response to the needs of growing urban populations and, perhaps indirectly, to the presence of a large army (Gaffney & Tingle 1989, 241). The early Roman period would have brought major demographic change less by force, than by the promise of more affluent lifestyles for an élite (probably comprising the leading local families from before the occupation) who could organise the necessary labour force. Dating of the manuring scatters suggests that it was during the first two centuries of Roman rule that cultivation of the estate was at its most intensive, and only towards the end of this period that the villa was constructed.

The later period proved elusive. Modifications of the villa and continued use of settlements suggested that this was a time of continuing prosperity, although a reduction by as much as 70% in the off-site pottery of this period might imply that intensive use of the land during the earlier part of the period had exhausted the soils, causing falling arable productivity (Gaffney & Tingle 1989, 240).

Gaffney and Tingle go on to estimate population figures and productivity for the Berkshire Downs, suggesting that the area would have supported between 500 and 1800 people, and that it would have generated a surplus meeting the needs of upto 2000 people. Due to the abundance of class A data a full narrative sequence has been employed, culminating in an elaborate but well founded class F economic narrative, which coheres well with existing general class G narratives, and allows the discrimination of hierarchy which constitute the bones of a class G narrative at local level.

In comparing this project with the East Hampshire survey a number of points should be made. Shennan's field team varied, starting with local amateurs and then making extensive use of the Job Creation scheme, as well as students. At Maddle Farm there were three full-time staff, supported by postgraduate students. Where Shennan has invested more energy in sophisticated statistical analysis, Gaffney and Tingle have sought greater spatial resolution through intensive fieldwork as part of a multi-layered approach which includes excavation. They have also paid more attention to the qualities of the artefacts themselves (in particular analysing the forms and spatial distributions of worn and broken scrapers compared with unbroken ones; Gaffney & Tingle 1989, 46-48), despite choosing to avoid the widely used length:breadth ratios as a means for assisting chronological diagnosis.

For the Romano-British period the benefits of the multilayered approach are pronounced to a dramatic degree. Although the greater proliferation of surface finds facilitated discrimination of land use, the particularising of detail achieved by excavation and test pitting within a core area provided a much stronger foundation for arguments concerned with distinguishing between settlement and concentrations which are off-site. Quite simply, the Maddle Farm project established a much clearer relationship between the surface and sub-surface.

4.3iiic   Tarragona, Catalunya

It was pointed out above (*4.3iiib*) that the volume of finds was a determining factor in enabling Gaffney and Tingle to perform their spectacular interpretative analysis of Maddle Farm. Off-site scatters in the core area reached an uncorrected mean of 4.1 sherds/ha. On the face of it, then, a survey which recovers sherds at a rate of 83/ha (Carreté et al. 1995, 53) should be well placed to articulate an extremely elaborate narrative.

The Tarragona survey set out to address questions relating to the influence upon settlement distribution of the major coastal settlement, of other local centres, and of natural features of the landscape. Four 1km wide east-west transects of varying lengths were set at 5km intervals across the north-south flowing Francolì, the southern most of which was 4km north of the city. The project report used here, based on the analysis of periods dated by ceramics, has been published separately from that of periods dated by lithics. Unfortunately, the latter publication has yet to appear, but in the mean time the former can be measured against the Maddle Farm analysis of the Romano-British period. Although the survey attempted some interpretation of land use, it was still subject to the allure of the site. Where it differed from most other surveys was in the rigour of its definitions. In the first place (see *4.2* above) it was recognised that the simple lumping of undiagnosed coarse wares with a few well known fine wares allowed surveyors to substantiate the existence of a site despite the lack of proof of contemporaneity from the finds. In the second place it was decided that the use of absolute numbers of finds to make period by period comparisons was inappropriate since it took no account of variations in rates of supply through time. Instead analysis was in terms of period by period proportional distribution.

When presented in period specific bar chart form it is quite apparent that the number of sherds per hectare in each field do not follow a statistically normal distribution (Carreté et al. 1995, 58-59; fig. 4.2) and tend towards bimodality, with the greatest peak in the lower ranges, and a second much lower peak in the highest ranges. Uses of the distribution mean would not represent this bimodality, so it was considered more appropriate to use the median collection unit (Carreté et al. 1995, 57). All the fields were ranked on the basis of the number of sherds per hectare within them, giving the 543rd of 1085 as the median. In the early stages of analysis quartile and octile ranges were used (on the basis of rank order allocating fields to four, then eight groups of equal size; Keay & Millett 1991, 137), but later it was found necessary to use 10 percent ranges (Carreté et al. 1995, 57) to accommodate the very low values from one of the four broad chronological bands, the Late Empire phase.

The amount of surface pottery varied sharply from phase to phase, but the median for all phases had a value of 0. The lowest positive value for the Iberian period occurred in the 40 percentile (i.e. within the set of fields ranked in the top 40%), for the Republican and Early Empire in the 30 percentile, and only in the 10 percentile for the Late Empire phase (where the mean for the 10 percentile remained at 0 to one decimal place). "[0]ne or more adjacent fields in the top 10 percentile" were designated to be ADABS. Although the authors allow that a variety of processes may have caused this distribution pattern it was noted that the application of the term sites to these concentrations was "supported either by the observations of remains such as floors and walls within exposed sections or through testing by geophysical survey" (Carreté et al. 1995, 61). However, it was noted that when density was as low as that for the Late Empire it was hard to distinguish between a site and a casual loss. Conversely, whilst the proliferation of sherds in the upper range of the Iberian phase may be regarded as a reliable indicator of the location of settlement or domestic waste, the exclusion of other findspots from the ADABS designation may obscure what were settlements which, perhaps of low status, had no easy access to pottery (Carreté et al. 1995, 61-62).

In a chapter entitled "The Identification of Sites" (Carreté et al. 1995, 166-214) each walked field was represented within a transect block for each of the four phases (Iberian, Republican, Early Empire and Late Empire), and shaded according to the percentile range of sherds recovered from it. The bulk of the chapter is taken up with what might be described as a gazetteer of sites, or ADABS, which has included more highly resolved chronological information, sometimes specifying evidence of activity within a specified part of a century.

For the purposes of analysis it was assumed "that surface scatters reflect the existence of buried remains" (Carreté et al. 1995, 216). Of the 10 percentile ADABS four were selected for more detailed fieldwork, testing the relationship between the surface and sub-surface. The sites were selected to represent "much of the topographical, chronological and size range of field scatters" (Carreté et al. 1995, 224), and they were each tested by non-intrusive geophysical technique's. Although magnetometry did not fair well in the particular conditions, resistivity demonstrated the likelihood that structural remains were present at all four sites. Geomorphological evaluation showed that no more than the floor levels and foundations were likely to remain at two of the sites, while some intact strata survived at the other two (Carreté et al. 1995, 223-230).

The selection included: Site 1.14 — four small terraced fields facing east-south-east, one of which produced high densities for all periods, one of which had high densities from the Iberian, Republican and Early Imperial phase, and one from the Late Imperial phase. Site 2.9 — a group of 9 fields "covering the crest and sides of a bluff overlooking the main valley and a tributary running ESE and SE" (Carreté et al. 1995, 225) where especially high densities from all periods except the Late Imperial were noted, the latter producing just a few sherds. Site 4.1 — two fields on a slight south-facing slope on a plateau with some Iberian pottery, but a stronger presence of Republican and Early Imperial wares (Carreté et al. 1995, 201 & 227). Site 4.6 — a single field including a probable enclosure of Iberian/Republican date. No topographic information is given.

A further site, found fortuitously in the section formed by a cutting for the new Vallmoll by-pass, was subjected to more intensive treatment, despite falling outside the four transects. The single field of approximately 0.2ha was divided into 5m squares, each of which was "thoroughly searched by a single walker". From the absolute number of finds per period, rates per hectare were calculated and adjusted to allow comparison with line walking figures. The surface material indicated ADABS for the Republican, Early and Late Imperial phases. Despite a mantle over the tops of features of 0.9 to 1.5m finds from the road section were in broad agreement with the surface pattern, and were associated with floor and wall features. Resistivity survey indicated several other structures set within one general alignment (Carreté et al. 1995, 216-223).

This is a sub-surface survey somewhat at variance with that adopted at Maddle Farm; no direct link has been established between surface artefact densities and geophysical anomalies. Plucking finds from a single vertical plane (the road section) below a line (where a collection was, if made at all, unlikely to be representative due to extreme recent disturbance during road construction) is no substitute for the coupling of intensity with wider coverage that can be achieved by test pitting. On the other sites, where there have been no attempts at all to investigate the relationship between surface and sub-surface finds, the use made of geophysical prospection is methodologically dubious. The principal need is for a comparison of like with like - of sherds with sherds - before any attempt is made to link surface finds with structures whose relationships with artefacts are unknown. Once such relationships from several examples have been investigated it may be possible to make suggestions about the implications that the distribution of surface material may have for structures or less readily perceptible residues of activity.

Failure to carry out a sufficient number, or the appropriate form, of tests does not negate the value of a narrative derived from surface survey, it is merely that the foundations for a hypothesis are weaker; the hypothesis can be tested at a later date. In other respects, the volume of the collected material from Tarragona and the attention to its composition offer potential for both chronological and functional resolution, although the extent of concentrations was treated as the primary means for classification of sites (Carreté et al. 1995, 231). Sites were divided into five size ranges: A - <1ha; B - 1-2ha; C 2-4ha; D - 4-8ha; E >8ha; the scheme was applied to areas actually walked, so if fields adjacent to a concentration had not been walked no allowance was made for this lack in the subsequent analysis. This was preferred to classification by the presence of architectural materials such as tile and flooring (*opus signum*) because, whilst they may allow differentiation of function and status, the former appeared only to be associated with post-Republican sites, and the latter was not datable. Elsewhere large storage vessels (*dolia*) have been taken as indicative of agricultural production, particularly wine and olive oil, but because at present there is no means of dating them, they can only be used where other finds occur. An advantage of determining site status by criteria other than portable artefacts is that the spread of valued artefacts, especially imported fine wares, becomes visible across the hierarchical spectrum. Whilst most or all high status settlements sites will attract certain prestigious artefacts, a few low status settlements may acquire them also. In this way it is possible to gauge the exclusivity of class/cultural discourse.

Once the general classification by size had been introduced and set against the four broad band chronologies architectural evidence was introduced as a secondary system of classification. It was assumed that if none of tile, *opus signum*, or *dolia* were present the site was of Low status. If one of these materials occurred (at least two sherds of *dolia*) the site was designated as Intermediate status. If more than one of these materials occurred the site was raised to Higher status. A fourth "Highest status" class, comprising a "clearly substantial villa," was not encountered during the survey, but examples were known in the area as a result of earlier work (Carreté et al. 1995, 235).

From a table making use of these classifications it was noted that "there is a clear correlation between increased size and enhanced status" (Carreté et al. 1995, 235 & 234; table 8.1). Because geophysical survey suggested that "relatively insubstantial sites" could be located from surface concentrations of limited extent it was felt "that the classificatory system was sufficiently substantiated to provide the basis for the analysis of settlement patterns. It had already been observed that the low incidence of pottery might be a function of its availability, rather than offering cogent evidence for a reduction in the size of the population and that it is impossible to comment on aceramic pre-Iberian and post-Roman phases (Carreté et al. 1995, 241).

The next stage of analysis looked at the relationships between the presumed settlements and landscape characteristics. Mapped geological information existed for the whole region (Carreté et al. 1995, 39-44; fig. 3.4), but there were gaps in the soils record (Carreté et al. 1995, 43-44; fig. 3.5). Existing topographical information on 1:5000 maps was supplemented by records kept during the surveying of each field. Fifteen geological units were distinguished within the region, of which all but four were walked. It was estimated that between 4.77 and 384.77ha of the other units were sampled, and from this it was possible to calculate the number of hectares walked before a site was found. Although this imbalance of coverage, and hence of sample size across strata, would weaken hypotheses concerning specific geological types a general trend showed that older rock and calcareous crust tended to be avoided, whilst "there was a more or less uniformly high density of occupation over most of the low-lying areas within the transects" (Carreté et al. 1995, 243-244). There was a similar correspondence between soil types and site distribution, with a some evidence for situations at ecological interfaces. There appears to be more distinct variation according to topography, which shows "a clear preference for dominant positions in the landscape" (Carreté et al. 1995, 246), although Allen has demonstrated the dangers inherent in the assumption that such a distribution is a consequence of discard rather than subsequent erosion (see *4.3i above*).

It may be acceptable to assess the degree of continuity on recognised sites (Carreté et al. 1995, 247; fig. 9.5) but the postulation of frequency of settlement and, hence, the extent of estates and village lands must be treated as weak for the reasons given above. Nonetheless, general observations concerning the distribution of small settlements with respect to larger settlements, with the implications for economic influence and affiliation, are more likely to be justified. However, changes over time in site sizes must be regarded with suspicion, not only because no allowance was made for ADABS where adjacent fields were not surveyed, but also because of the breadth of chronological bands. Equally, inconsistent modes of data collection introduced from earlier work to fill the gaps between transects undermine the credibility of assertions about the relationship between the density of sites and the distance from Tarragona (Carreté et al. 1995, 251-252; fig. 9.7).

One of the most successful features from the analysis is that of the distribution in time and space of pottery, heralded elsewhere well before the appearance of the final report (Millett 1991; Keay & Millett 1991), but given in more detail in it. The general method is a simple but effective one which might be adopted elsewhere, using the most frequently occurring fabrics from each phase, and deserves to be set out in full (Table 4.2):

Table 4.2 — Percentile analysis of ceramic distribution (Tarragona)

1   Establish the total number of sherds of each fabric within the whole survey area

2   Establish the percentage represented by each fabric within

    (a) the total survey assemblage, and

    (b) the assemblage from each ceramic phase for the whole survey

3   Total the sherd numbers for each of the fields within the site

4   Establish the percentage represented by each fabric within

    (a) the whole site, and

    (b) the ceramic phase to which it belongs on that site

5   Compare the percentage of each fabric on the site (from 4) with the equivalent value for the whole survey (from 2) (Carreté et al. 1995, 253; table 9.5)

From this approach it was possible to note the penetration of imported vessels into systems of local production, and their effect on the continuity of production over time, as well as providing evidence for the degree of localisation of ceramic production. In some cases, of course, where it is assumed that substances contained in vessels were the imported matter, rather than the vessels themselves, it was possible to say more about the nature of exchange networks. Data are presented in a series of bar graphs which illustrate each of the chosen fabrics' deviation from the survey mean on a site-by site basis (Carreté et al. 1995, 254-264). This level of interpretative analysis, which lies in classes E and F, was then introduced into hypotheses about settlement size and status to give further substance to class G discussions of settlement hierarchy.

The most disappointing aspect of the report is its failure to generate substantive hypotheses for off-site distribution. The authors merely allude to accidental and incidental discard, and refer to manuring without attempting analysis. Aware of the work at Maddle Farm, they argued that such an attempt would be undermined by the surfeit of surface material, and complain that the temporal span they are dealing with is too great, causing blurring of discard patterns (Carreté et al. 1995, 266). This is a curious argument for a project which has been at pains to give more than usual attention to chronology. The Romano-British period was broken down into two phases, the earlier covering around 150 years, the later around 200 years. This resolution is little different from the four phases used at Tarragona, which had durations of between 200 and 300 years.

To summarise: the techniques used in the field were not sufficient to sustain strong arguments about particular settlement density, but were capable of showing at a general level the changing relationship between the rural hinterland, the coastal urban centre and the Mediterranean macro-region over a one thousand year period. Within the region it was possible to infer distinct localities of influence which were more pronounced at different times. The arbitrary use of the highest 10 percentile of field densities to diagnose an ADABS or site is both a strength and a weakness. In its favour is the construction of distinct entities for analysis and comparison; against it, in this particular case, is the wholly inadequate treatment of off-site data.

4.3iiid   Thy, Jutland

The Thy Archaeological Project combined a keen knowledge of the literature with labour intensive field procedures to generate data which were interpolated to give total representation of densities within a given area. This is the kind of "multiplying up" which Foley used to estimate the volume of lithic material likely to have been discarded in the Amboseli, but in this case it is first applied to perceived discrete units which have been conceived of as sites, and then to amplify secondary locations concealed within the background noise. In particular, for his theoretical framework, John Steinberg borrowed, and redefined for his own use, the term "signature" (Schofield 1991b, 5):

> "... we quantify broad categories of artefacts and calculate their frequency over a site, which gives us the raw stuff of a site signature. The goal is to understand how the amount and efficiency of lithic production correspond with use of stone tools" (Steinberg 1996, 375).

Steinberg is asserting that the ploughsoil, rather than the pattern of largely ploughed-out subsoil features, actually is the greater part of the *site*, an observation made by other fieldworkers at least as early as 1980 (Yorston et al. 1990. 68). The ploughzone is represented as a homogenous fluid within which artefacts move according to particular and general constraints imposed by cultivation, topography and erosion.

The field procedures for the project have been discussed above (see *3.3ixc*). Obviously, as a regional procedure (which the author describes as "intensive"; Steinberg 1996, 385) it has limitations as it takes up to half an hour to process each 400 litre sample using a purpose-built machine, with a minimum of twelve samples per hectare taking upto a day, and with on site intensification taking upto five days to complete (and that an artificial and arbitrarily imposed upper limit). By reference to the literature on experimental uses of seeded artefacts it was calculated that "samples placed any closer than 4 m apart will be redundant. Samples placed further than 20 m apart have a distinct possibility of missing concentrations" (Steinberg 1996, 374).

It was suggested that a low frequency of sampling would increase resolution by producing steeper gradients between collection units. Secondary and tertiary sampling stages allowed intensification of work where higher densities occurred, so generating the potential for more meaningful quantitative analysis. In addition to this intensive procedure some shovel testing was carried out, but on a scale which adds little to the arguments in the present paper (Steinberg 1996, 385).

The project sought to distinguish between sites (it is not clear on what numerical basis a site was recognised) by period and function. In Britain it is becoming standard practice to calculate length:breadth ratios of flakes as a means for dating, along with detailed investigation of the mode of knapping (for instance, Phil Harding in Woodward 1991, 74-86). At Thy this option does not seem to have been considered; instead, after testing chronological variation according to flake size (presumably, the longest dimension) at two sites it was decided that weight correlated closely with size, and provided an approximate means for diagnosis (Steinberg 1996, 375-376. Greater average weight of flake was generally indicative of an earlier date for a site). For the purpose of functional analysis flint flakes were classified according to whether they had been retouched, burnt, were whole or broken, or whether the flint was a "chunk". It was assumed that a high proportion of retouched flakes was a good indication of consumption, whilst the "raw number of whole flakes" was "a good index of production" and chunks were "an unequivocal indicator of lithic production" (Steinberg 1996, 377).

The aim of the subsequent analysis was to calculate a pseudo assemblage for the site by interpolation. A Thiessen polygon was drawn around each 400ml sample and, on the basis of the surface area represented by that amount of soil (making allowances for variations of plough depth around an estimated average of 30cm) the number of all tools and each flake type per sqm was calculated and multiplied by the number of sqm enclosed by the boundary of the polygon. From this it was possible to estimate the percentage of retouched tools, and the number of "waste" flakes per different categories of tool (Steinberg 1996, 380; table 3). Once tabulated it is possible to read site signatures with distinctive patterns of discarded material which form the basis for functional and, to a lesser extent, chronological discrimination.

Three sites were selected for special scrutiny to illustrate the advantages of classifying and interpolating data in this manner. After converting the volumetric flake incidence rate into an average rate per square metre in the manner described above sites THY 2788 (Early Bronze Age), 2578 (Late Neolithic) and 2981 (Early Neolithic; were assigned estimated total flake populations of 237,300 (over 13,000sqm), 24,700 (over 1600sqm) and 666,300 (over 36,300sqm) within calculated ranges at a 95 confidence interval. This statistical procedure (described by Dwight W. Read; Steinberg 1996, 391-392) assumed a random distribution of artefacts sampled within an arbitrary pattern which tended to "minimize any spatial autocorrelation effects". By the same means collection composition was represented in greater detail (Steinberg 1996, 381; table 4), including the different recognised classes of tool. THY 2788 was calculated to have 1910 tools against 2578's comparable 1540. At first sight this might be taken to imply that tool production at 2758, generating only 16 flakes per tool, was much more efficient than at 2788, where 124 flakes were generated per tool. However, it is suggested that 2578 ought to be treated as an "end-user" site (a site where tools are consumed, rather than produced) since no cores were found there, whereas tool production at 2788 constituted an *ad hoc* meeting of daily needs. THY 2981 represents a third kind of site where production was specialised. Although the number of flakes and tools is much higher than on the other two sites the rate of flake to tool was three times less than at 2788, but two and a half times more than 2578. Amongst the tool class were a calculated 3,560 polished axe fragments which, it was thought, might represent some 350 axes.

By generating figures for the total volume of lithic artefacts rather than merely comparing directly observed rates of recovery for each artefact it was possible to make inferences from the varying proportions of discarded material which were tempered by estimated periods of occupation. On the basis of previous studies made of Danish prehistoric architecture it was assumed that "all sites were occupied year-round for 10 years" (Steinberg 1996, 382). With this in mind, THY 2788, was estimated to have generated an average 65 flakes each day, and a tool discarded every second day. At THY 2578 the average daily rate of flake production would have been only 6.6, although the number of tools predicted was not significantly different from 2788. Both sites were situated within easy reach of moraine flints, but it would appear that although tools were discarded at a similar rate they were not being produced at equal rates on the two sites, thus supporting the argument that one site was a lithic production centre, while the other consumed tools.

The numerical specificity of the argument may appear as a strength in this prize-winning paper (Chippendale 1997, 1), but the gap between the raw data and the estimated figures strains credulity. The line of argument depends upon several assumptions, including: that the soil is truly well mixed, allowing a representative sample to be gained from comparatively few test pits; that a significant degree of equilibrium of displacement by ploughing will occur; that the finds from each site represent a single phase of occupation.

The notion of soil homogeneity, where artefacts are part of the composition, depends on there being a substantial number of each particular class of objects which is subjected to statistical procedures. At THY 2981 an average of 18 flakes/m2 and a high of 118 flakes/m2 provides an ample basis for analysis of flake distribution from 37 different 400ml samples, but a mere 10 fragments of polished axe do not. It might be helpful to know their horizontal distribution to see if there is a particular concentration, but there are no grounds for multiplying up from so small a sample, and basing higher level interpretations upon a figure of 3500 fragments (Steinberg 1996, 384). Unfortunately, because raw data figures are not given for other tool fragment categories it is harder to judge the appropriateness of the figures (Steinberg 1996, 381; table 4), although if the estimated figure of 8200 scrapers is divided by 350 (the number by which the actual incidence of axe fragments was multiplied) a total of around 23 scrapers from 37 samples would not represent a firm basis for judging the total population. If Steinberg is arguing that discrete sites are represented by concentrations, why not localised deposits within the sites of particular artefacts?

This last point is a feature in arguments about equilibrium of artefact movement. Equilibrium does not mean no movement; it is a stage where repeated ploughing should bring about a "constant average displacement" (Steinberg 1996, 373), although that average will vary with the shape, density and size of an artefact (Clark & Schofield 1991, 96-99). This variation, where objects have a single source, should result in an increasingly marked differentiation between, for instance, cores and flakes, or flakes of different weights. Steinberg suggested the higher average weight of flakes in soil lying over an Early Bronze Age structure compared with those outside it might be a reflection of equilibrium patterning. However, the same experimental evidence undermines the significance of the correlation of core and flake distribution, despite the strong positive indications resulting from regression analysis ($r^2 = 0.61$).

The third issue is the thorny one of contemporaneity. Each of the three concentrations have been treated as though the products of a single phase; more specifically, Steinberg has suggested each phase might be of approximately ten years duration. Although the paper suggests that it is possible to distinguish between phases according to the weight of the flakes, this is conducted as an average of perceived discrete site collection; variation of date within the perceived site is simply not addressed in this paper. Most remarkably at THY 2578, despite strong evidence for overlapping structures of similar construction following the stripping of the ploughsoil, analysis of discard rates is still based on a single ten year period. Some of the project's class F conclusions may be justified, but the failure to make detailed comparisons between broadly contemporary sites is a weakness in this regional survey which will no doubt be addressed in a later report. As an example of field technique and sampling the Steinberg's paper is of considerable interest, as are the theoretical constructs underlying their use. The single greatest problem in the report is the manner of statistical inference. The high density of lithic artefacts generally does indeed legitimate some of the bold assertions about the volume of flints within the ploughsoil of a given area, but the sample size of most of the specified tools is too small to be mitigated even by a volumetric approach, especially where artefactual chronology is poorly defined.

4.3iv The role of statistics in regional archaeology

In sections *4.3i* and *4.3iii* I have presented examples of different statistical methods by which the archaeologist may test whether or not the relationship between variables is significant, and how strongly so. Substantial portions of these analyses are concerned with the relationship between the distribution of artefacts and present or, where they can be reconstructed, past landscapes, sometimes making allowances for the influence of the raw data collector. They seek to show whether or not activity areas are influenced significantly by topography, soils, ecology, the presence of water etc. (class E), and may then move on to explanations which fall within the socioeconomic sphere (class F). We have seen that to a considerable degree spatial resolution can be controlled by decisions about sampling, and that in most regions means can be found for overcoming, or making allowances for, the various problems of visibility.

The single biggest problem remains the achievement of a satisfactory level of chronological resolution, without which contemporaneity cannot be established, so rendering accounts of inter, and even intra, activity area relationships highly speculative. There is demonstrable potential for the addressing of such issues in some periods, but a reversal of approach may be required. Where previously the favoured object's of Mediterranean survey were fine ceramics, it is increasingly obvious that it is possible to describe coarseware chronotypologies. Imported and fine wares tend not only to be rarer than local fabrics, but they may well have a longer life. Where finewares may prove useful is in building an associative sequence from stratified contexts within a study region which can be applied to surface or ploughzone finds. In southern Britain, for instance, detailed knowledge of Dorset coarsewares (Black Burnished Ware, BB1) excavated at Dorchester provides a detailed sequence for a pottery industry which often accounts for more than 80% of the material from Romano-British sites in the southwest counties. This sturdy fabric will always be represented in surface scatters of that period in the region.

In the foreseeable future, surveys targeting the Romano-British period are likely to introduce distribution plots broken down into periods spanning just a few decades, showing the probability of chronological overlap between various activity areas. The same methods might be applied in broader bands to the later Prehistoric periods, although with a low survival rate the ceramics are likely to be of dubious statistical value, unless more archaeologists are willing to attribute undue significance to ADABS figures as low as 1 sherd, as happened for the Late Imperial period at Tarragona.

Lithic industries continue to present a problem; reliable and detailed regional typological sequences have not been established. It would seem that researchers are to be condemned to broad banding in which, for instance, the European Mesolithic cannot always be distinguished from the Early Neolithic, and the Early Neolithic is not entirely visible against a Late Neolithic/Early Bronze Age background. It may be that close examination at local level of stratified lithic types associated with ceramics, and thoughtful presentation of the two together, will lead to better definition, but at present it is likely that the only honest way of judging the chronological range of a scatter is to sort and analyse all the diagnostic material, and consider the potential associations with it of undiagnostic material. Of course, it is important to avoid circularity; although a portion of the diagnostic material may belong to a certain period, it should not be assumed that a similar proportion of undiagnostic material also belongs to that period. Assumptions of that kind would entirely mask the distinction between a "producer" and a "consumer" concentration, for example.

It might be argued that the Thy project's use of statistical analysis opens a way forward to higher levels of interpretation at regional level. However, I would argue strongly that it is the choice of field techniques which has had the strongest influence over interpretation. Thy's techniques were indeed sensitive enough, and the sample sizes large enough, to pick up significant trends in the inter-site occurrence of particular artefacts. Nonetheless, the interpolated figures for low incidence items ought to be treated as spurious and misleading when introduced to support higher levels of argument, even before problems of collection/ assemblage contemporaneity are considered.

The only effective means to deal with this latter issue is the use of stratified data. Steinberg would have done well to look more closely at the relict features which topsoil removal showed to be under the ploughsoil. For example, comparatively few finds were recovered from the features at THY 2578, but they might have proved valuable comparanda for discriminating contemporary ploughzone material.

Earlier Prehistoric survey can either restrict itself to making very general statements about population movement and preferences over very large tracts of time (Foley 1981; Gaffney & Tingle 1989; Shennan 1985 etc.) or they must adopt much more intensive techniques at a few locations within a region, making it possible to make claims at the class G and enhance class F levels. It is sensitivity to place and to the location and characteristics of the population of artefacts, rather than statistical inference, which structures the narrative for the Dorset Cursus.

There is a danger that the sheen of numerical argument may obscure imagination and reasoning by word, when the categories upon which the former is based are no more secure than the conclusions of archaeologists pursuing a more intuitive approach to their subject.

4.4

Find: Fit: Fable — The limits of behaviour

> "If I find variability in patterned configurations, I have some evidence for dynamics, changes that occurred in the past. I know that something happened, that some dynamics were operative, but I do not know why the changes occurred; neither do I know anything about the character of the changes. To make a statement about the character of changes I must first assign meaning to the contemporary facts of the archaeological record." (Binford 1983a, 179)

Looking at a map on which the incidence of surface finds from a particular period has been plotted often leads the viewer to make the tacit assumption that he or she is looking at the distribution of co-existent settlement. Even those who know better frequently judge human population sizes and exchange systems by such maps. Colin Haselgrove recognised this problem over a decade ago, when considering the analysis of surface collections. He urged that "practitioners must face up to the interpretational challenge and to the necessity of having an inferential methodology specifically tailored to the nature of their material and its peculiar problems" (Haselgrove et al. 1985, 14). Ultimately, if regional survey is to make sustained progress beyond being the provider of elaborate site location maps more attention must be given to Clarke's fundamental unit, the artefact, but also to the changing entity into which it was incorporated, the landscape.

The artefact is always a passive expression of direct and indirect action. Nonetheless its form also influences the pattern of human behaviour both in its making, in its use and in its discard. Once discarded it may be consigned to a stage in its existence where it has minimal influence, although it may continue to be the subject of natural and human agency, whether locked in the medium of a stable soil deposit, temporarily placed in an agricultural manure dump, or moving downslope during severe erosional episodes, prior to secure burial in a valley bottom.

It is the distinctive task of the regional archaeologist to diagnose from these artefacts not only the activities associated with them but to distinguish places to which those activities were particular, and the significance of that particularity. In the following subsections I shall attempt to show how that can be achieved through understanding the constraints imposed by the artefacts form and distribution. Many of the issues bear a close resemblance to those arising from any attempt to generate an archaeological narrative.

4.4i   Perceptions of constraint

It is hard to imagine that only two decades have passed since Richard Hope Simpson stated that "... the limits inherent in all surface work demonstrate that the supposed superior value of the new "intensive" surveys has been much exaggerated" and went on to berate manual and computer inference as "almost wholly inapplicable to survey work" (Hope Simpson 1983, 45 & 47). The value of survey is no longer seriously doubted, and instead it has arrived at the sharp end of a continuing debate over archaeological interpretation, in which the participants occasionally change clothes. It is between those who see space as the container of human behaviour, with ecology as "the external setting of the system in its environment" (Clarke 1978, 101), and those who see a particular space, the landscape, as the medium conceived within human praxis (Tilley 1994, 10). Susan Alcock asserts that "The aim of archaeological surface survey is to locate and relate in a diachronic perspective all remains of human activity across a landscape; it thus operates on a broad temporal and regional scale" (Alcock 1993, 33). Thereafter it is sufficient to plot contemporary settlements, to characterise finds from them, and to hypothesise about exchange and influence of different polities. The authors would object to the claim, but the underlying perspective of the Neo-Thermal Dalmatia Project was similar. Although tradition was considered to be embedded in a landscape "in which pre-existing monuments are often used by later generations for their own distinctive purposes", their "key task" was to identify "the landscape structures defining each region (Chapman et al. 1996, 9-10). Those adopting the position of Tilley would view this as pseudo objectivism; their objective reality is that the landscape is conceived by each subject, within a grammar acquired in the course of social reproduction (Tilley 1994). In the Dalmatian project landscape has been reduced to land*form*.

In using GIS technology Marcos Llobera has adopted a phenomenological perspective, so emphasising the important distinction between region, a present day entity or arbitrarily designated research area, and landscape which is a matter of perspective, "the general idea of locating the centre of reference in a *mobile* individual's body rather than at a fixed point in our study area" (Llobera 1996, 613). As yet our technological limits cannot give us the full somatic experience, but GIS does at least allow modelling of the physical constraints on a subject's eye view of the land and its features, in this case allowing Llobera to consider the guiding principles and impact behind the layout of a group of Wessex linear ditches.

This subject-situated approach is very appealing, but if our goal is to grasp what it was to be an individual member of a prehistoric group we must not only understand the physical shape of the landform at the time, but must rely on more than intuition to understand how the human population was integrated into the landscape. It is difficult to avoid reverting to the concepts of analytical archaeology. Clarke proposed "a general model system" within which particular culture systems could be recognised and described. The characteristics of a system are that it is "dynamic and continuous, with the attributes or entities having specific value or states which vary by successive transformations" (Clarke 1978, 45). It is very much artefact based, and as a means of research selective, excluding data regarded as irrelevant, but dealing with essential and key attributes whose "values or states change as part of the changing system" (Clarke 1978, 75). The development of the attributes of an artefact, those characteristics of form and decoration by which a culture is recognised, vary and change within a trajectory (the dynamic potential of tradition); new modes are constrained by the knowledge of the existing and former modes.

By detailed application of this method Clarke hoped to achieve optimum phasing, and to define the extent of culture groups, including their expansion and contraction over time within and beyond a region. Although he has provided a language and methods of research, his teleological aspirations have been subject to much criticism. In reality the processual analytic approach has probably contributed much more by focusing not on direct explanations of human behaviour, or general theory, but on "middle-range theory". The attraction of middle-range theory as expounded by Louis Binford is that it looks at constraints upon actual behaviour which may be inferred from the artefact. In this sense it is pragmatic. The archaeological world can be divided into the directly perceivable statics ("information preserved in structured arrangements of matter"; Binford 1983a, 416) and dynamics, those processes of which the statics are the products and hence to which they testify. This is not a solution, merely an indication of what the researcher might need to look for.

Binford has suggested that the "energy" needs of cultural systems and their individual "participants" necessitate "conditioning interactions" with the specific environment, thus selecting "for and against certain culturally organised means of articulating" with that environment. "These points of articulation provide the determinant dynamics for descent with modification", they are the motor of Clarke's trajectory (Binford 1983a, 222). Points of articulation, those moments when in a Hegelian sense an individual or culture risks displaying its distinctiveness as well as its belongingness, may have had a form of which there is now no trace, such as utterance or dance, or may have had substance in the shape of the artefact. In this sense the static informs the dynamic. It is hard to conceive of a static, for instance a vessel, being the motor of change; rather it is the immediate form of what has changed, and a signpost to further change.

This has brought us back to the analytical starting point: the artefact. How are we to identify the "... dynamic conditions responsible for the statics" (Binford 1983a, 222-223)? Binford's answer is straightforward: "It is necessary to experience directly the process of adaptation and in turn the archaeological products of this process" (Binford 1983a, 191). The need is for "... actualistic studies designed to control for the relationship between dynamic properties of the past about which one seeks the knowledge and the static material properties common to the past and the present" (Binford 1983a, 421). Here, then, is the point of Binford's ethnoarchaeological work amongst the Nunamiut - the behaviour he observed while amongst this high latitude people can be applied to Mousterian hunter-gatherers. The assumption is that the urge to survive in a glacial climate, coupled with a similar technological level, will generate a broadly uniform pattern of behaviour in different times and places.

The uniformitarian approach, a method of explanation borrowed from geology, does not sit happily with the modern, particularly western, world's ideological plurality. Nonetheless, Bruce Trigger has suggested in his account of archaeological thought, that none of its opponents have "succeeded in producing a credible alternative". He goes on to bemoan the prevailing tendency merely to classify artefacts according to their date and culture, rather than in addition exploring the "significance of formal variation for understanding ecological, social, political, ethnic, symbolic, and ideational aspects of prehistoric cultures", as a "major technique for bridging the gap between the archaeological and behavioural spheres" (Trigger 1989, 366; see also *4.1* above). While interest in such aspects is certainly desirable, the issue of contemporaneity once again raises its head. Rosamund Cleal, analysing Neolithic pottery from Cranborne Chase in terms of traditional categories, observed that vessels associated with similar ^{14}C dates were unlikely to be the products of the same group of potters, but that it was at least as likely that those groups were separated by time, as by situation in space. However, extending beyond the Cranborne Chase region to a macro region inducing Wiltshire sites, Cleal was able to observe patterning in the distribution of particular Grooved Ware motifs, although this could not be shown to extend further afield (Barrett et al. 1991a, 141-143). But these patterns only start to make sense when John Barrett and Richard Bradley bring to bear their ideological perspectives upon them (Barrett et al. 1991). In short, it is when data are tested against a value system that meaning begins to emerge. With this in mind we should be looking to see what sorts of hypothetical expectations of past worlds might best be tested, even contradicted, by survey.

This should be regarded as a pragmatic approach. It expects survey archaeologists to combine their archaeological and ideological experiences to construct theories about the regions and periods they are about to investigate and then to strive to prove them. There is nothing new in this; Popper regarded it as a routine part of the pursuit of science, and in archaeology the likes of Schliemann, Childe, Clarke, Binford and Cunliffe have all ventured theories from the site to the continental level which have since been contradicted. The point is that precisely through their initially plausible propositions of what might be, contradictory information has been highlighted or sought, not only undermining part or all of a theory, but providing new constraints for any subsequent hypothesis.

I am arguing not simply that preconceptions ought to be a conscious part of project design, but that as we deal with data at the class A-E levels we should be considering the consequences of the data and its interpretative analysis for our preconceptions. If one of our hypotheses states that "Late Bronze Age burnt mounds are the remains of saunas" and we believe that saunas must have easy access to water, when we find a burnt mound without such access the next stage is not to relinquish the hypothesis, but consider ways in which changes in the local hydrology might be explored! If a general theory states that there was no Roman military occupation in Ireland and evidence for such an occupation comes to light, look closely at the chronology and consider whether or not it was an abortive military occupation! In this way we will indeed perceive contradiction and, with it, constraint.

So what are the sorts of constraints which might be offered by survey data? Clearly that will depend on the techniques employed, the intensity of their application, and the extent of the region researched. In the case of surface survey, the first step is to decide the frequency of collection and plotting. If there is a target period, this may be decided by experiments at a number of locations to determine a range of artefact sample sizes from lines or grids of varying proportions. If the aim is to establish a map of general density patterning for a period where artefacts are frequent and durable on the surface a small collection unit will suffice; in any event it should always be possible to group several collection units to form one sample point. The only advantage of using large collection units is time-saving. There is no reason in principle why the sample point's area of reference should not vary from period to period, allowing greater specificity for phases where material occurs more frequently.

4.4ii *Concrete thought: ritual and economy*

We have seen how data distributions can offer insights of a general kind into economic structures through statistical analyses in the cases of the Thy and Maddle Farm projects. Other surveys, which have been far less rigorous in their use of numbers (generally implying a less thorough approach to class E interpretation), have contrived more detailed narratives at the class F level, subsequently used to develop class G themes. Usually, these accounts are derived from "site" or findspot gazetteers which either remain unpublished, or appear in the form of microfiches. A few reach the printed page, but in crude, summary form (for -instance, Lobb & Rose 1996) which makes the inferential link between data and higher level narrative tenuous. Notable exceptions are the detailed interpretative site gazetteers from the Biferno Valley (Barker 1995a, 1-35) and Sergemes Valley projects (Dietz et al. 1995, 134-175), and the East Anglian surveys (Hayes & Lane 1992; Lane 1992; Hall 1996 etc.) which might be regarded as the apotheosis of the genre. Very often class D and E narratives are compressed into such gazetteers. Some are accompanied by maps, others merely provide map references, with some indication of the area included within the site.

4.4iia Sergemes Valley, Northern Tunisia

The archaeological survey of the Sergemes Valley project was founded upon detailed geological and environmental research. Drastic erosion had at one time created deep blankets over sites, then carved out sections revealing profiles of buried buildings and the strata burying them. Because of the exposed structural remains there was a strong architectural component to the survey, and this was reflected in the overall objectives of the survey. It was hoped to provide a detailed map of Roman settlement in the context of particular functions, the roads and centuriation, water resources, agriculture, access to raw materials and industry. Twelve localities, of areas varying from less than 2sq km to 9sq km, were selected to represent the topographical variation within the study area with a view to investigating the relationship between a core area (around Sergemes itself) and its rural hinterland (Dietz et al. 1995, 116-118). A basic method of evaluating the quality of information from the site, based on the association of finds with structural remains, is set out in Table 4.3:

A	Datable structures of identified function.
B	Datable structures of unknown function.
C	Undatable structures of known function.
D	Undatable structures of unknown function....
E	Areas without structures....
F	Other types of finds apart from structures and ceramics, for instance data from prehistoric periods....
1	High concentration of sherds, many datable finds.
2	High concentration of sherds, few datable finds.
3	Low concentration of sherds, many datable finds.
4	Low concentration of finds, few datable finds. (Dietz et al. 1995, 125-126)

Table 4.3. Classification of site remains (Africa Proconsularis)

A straightforward numerical definition of sherd density was used, comprising the categories: 0, 1-10, 11-20, 21-50, 51-100 and greater than 100 per 100 x 1m line (Dietz et al. 1995, 122). Each fieldwalker carried a counting device to give a cumulative record of the number of sherds in every 100m line (see *3.3ixd*), whilst "representative sherds" were collected (Dietz et al. 1995, 122). Combining these data with the structural record it was possible to classify sites by a simple letter/number combination. Thus "undatable structures of unknown function" associated with "high concentration of sherds, few datable finds" would be described as D2. In fact, no A or B classes were found during the survey (Dietz et al. 1995, 131; 175).

Maps of each area detailed where fieldwalking had taken place and the density of surface material by line, generating greater spatial resolution than is generally the case. Unfortunately, there are no chronological divisions, so greatly undermining the value of the maps. Each gazetteer entry includes description of the surface condition and topographic details, as well as an estimated size of the field, the number of sherds encountered, and the range of sherd rates per line. Finds and architectural features of special interest were also described in terms of date and function where possible (Dietz et al. 1995, 134-175). The length of a separate chapter dealing with the architecture in more detail (Dietz et al. 1995, 179-347) emphasises the importance of this aspect to the project (*3.3v*, above).

From this information, which was much more detailed than that available from most regional surveys, it proved possible to diagnose sepulchres, fortifications and habitations. For instance, where ceramic evidence was lacking for later periods of occupation studies of architectural material strongly suggested that there had been continuity or re-use. However, this survey suffers from the same complaint as many Mediterranean surveys: an over-selective approach to collection, coupled with a narrow diagnostic range. Only fine wares were used for establishing site chronologies, leading to the absurd situation where, although there was a superabundance of material, a commendably focused sequence divided into blocks of fifty years was supported by "1" or "more than 1" sherd per block (Dietz et al. 1995a, 776-780). It was assumed, based on the very low incidence of pre-Roman finds evident from the gazetteer, that the valley was "a culture landscape created by Roman/Vandal and Byzantine civilizations" but that the region had been modified drastically by modern human agency (Dietz et al. 1995a, 773). A further conscious bias was introduced by the practice, after the first season, of working much more intensively where there were architectural remains and, consequently, greatly increasing the count in their vicinities.

The advantage to the reader of this format is that it is easy to make one's own basic assessment of the evidence. When Dietz asserts that for sites "without recognizable architectural structures, five sherd concentrations probably represent dwellings" (Dietz et al. 1995a, 774) it is easy to turn to some of the specific examples he cites. In area 1 sites 14-17 range from category D4 to E3 (Dietz et al. 1995, 138), the last three being areas without architectural material. Two are categorized as high density because they include 100sqm lines with a sherd count exceeding 100. All include some line sherd densities which easily exceed five, including a range of finewares, in areas of above average visibility (75-100%). Referring to the association of such finds with structures in other parts of the survey region Dietz asserts that buildings, specifically dwellings, would have stood on these sites. This seems perfectly plausible, although alternative suggestions such as midden dumps ought not to be excluded. Much more tenuous were observations of site 6, where large hewn limestone blocks and wall fragments, associated with a mere four sherds, were judged to have "fallen from above", although, it was noted, they could have been moved from elsewhere (Dietz et al. 1995, 136).

On this analysis, the frequency of sites during the period from 0 to AD700 varied between one to two per sq km, but again it should be stressed that without greater chronological support the density figures should be treated as an unreliable. As a presentational ploy Dietz's "first use diagram" (Figure 4.8) has much to commend it, but it would have been much more compelling if he had drawn upon a wider range of wares to support it. The diagram shows all the sites which produced dating evidence arranged in descending order based on the earliest incidence of more than one sherd of diagnostic fine ware in a 50 year block. These blocks are represented by a black fill. The more tentatively dated blocks with a grey fill produced only one such sherd (Dietz et al. 1995a, 777-778; figs. 2 and 3). Span of occupation is suggested by subsequent black and grey-filled blocks.

Derived from this information, summarised from the gazetteer, at least seven Punic settlements appeared to have been founded on land lying above 100m asl, and one on the floodplain; then, after a hiatus which only two survived, there was tentative settlement growth, becoming more rapid, in the first century AD, tending towards lower lying slopes above seasonal river courses (*oueds*). The pattern of growth was illustrated with a series of small scale contour maps, representing each 50 year block of time, on which 1km grid squares were filled in where settlement was recognised within them. After a summary map of Punic settlement, fourteen further maps cover the period from AD1 to AD700. This attempt at addressing the issue of contemporaneity (Figure 4.3, above) is superficially appealing. It does show marked patterning in both the density and the distribution of settlement. In early periods of low density, distribution tends to favour the valley sides; during the 150 years after AD400 the valley floor is equally covered, and as settlement decreases thereafter visible settlement becomes rare and dispersed, again favouring higher locations. The problem with the maps lies in the lack of discrimination of density within the 1km grid squares. Although it is possible to make out a trend of focal activity around Sergemes from the fourth century, after it been designated *municipium*, on the basis of careful reading of the gazetteer, the maps suggest that there is little to distinguish it from areas on the north east side of the valley, or even the valley floor areas. Representation of varying density of occupation per grid square would have done much to enhance the perception of difference and its significance in the wider economic scheme. From the gazetteer it proved possible to distinguish a seven-tiered site hierarchy, with time-specific variations in settlement type, density and location (Dietz et al. 1995a, 789-795). However, it is noticeable from the "First use" diagrams that in periods of denser settlement during the fifth century there is a marked reduction in the proportion of diagnosed smaller scale settlements to larger scale settlements. If scale of settlement was taken as an indicator of status or focalness for fine ware trade, this might simply be a result of a failure to find fine wares at some sites. The inclusion of "undefinable habitations", if taken to represent villas and farms, would largely make up this shortfall (Dietz et al. 1995a, 794-795; fig. 12), a possibility made more substantial when linked to the chronological distribution of agricultural implements found in proximity to diagnostic sherds.

Finds from ten sondages were checked against the material observed on the surface, eight of which were situated at sites associated with a structure, two of which tested surface concentrations. These demonstrated a rough correlation between the horizontal and vertical dimensions, although not all phases represented on the surface were present in the stratified deposits; the reverse was also true. Only in two cases was there "total correspondence" (Dietz et al. 1995a, 786-788).

Despite access to good quality environmental data, in the pollen sequences of which wheat and barely featured much more strongly than olives, there was no attempt to reconstruct the economy beyond reference to historically testified events over the period studied. There is a thin class F narrative, and a veneer of class G, but the great mass of data retrieved from the field has been much underused, and the reader is left to draw his or her own conclusions! The authors cite the lack of firm chronological data as a reason for their cautious approach to narrative.

4.4iib The Fenland Project, East Anglia

In terms of scale, the most ambitious programme of fieldwalking in Britain was that conducted in East Anglia during the late 1970s and 1980s, which remains in the process of publication as a series of monographs. Between 1985 and 1998 eleven volumes had appeared as part of the Fenland Project (Hall 1996, 230). The project's aims have been summarised as "a large-scale archaeological and environmental reconnaissance survey" of an area where changes in agricultural practice are threatening an especially rich resource, considered to be of international importance (Hayes & Lane 1992, 7). The nature of the landscape remains one of continual fluctuation, dependant on its interaction with the sea. This has had a determining impact on human settlement and activity, which has also ebbed and flowed according to whether contemporary technologies have allowed exploitation of a marginal natural resource, or whether particular general "needs" have presented the opportunity for the industrialised exploitation of resources such as salt.

This is a paradoxical landscape where silted watercourses (roddons) have provided ridges and islands which stand proud of marine clays, exposed by prehistoric and later peat depletion, and offering soils on which to build and farm; yet it is a region where much Bronze Age archaeology remains concealed by flood and marine deposits. As such it is a land of great potential for researchers, but with demanding though well understood problems to overcome. The environmental data from the Fens appear to offer clear boundaries of constraint in terms of the availability of land for settlement, as well as of the potential for a limited range of means for subsistence. It is tempting to believe that once the different phases of inundation and their effects upon soil have been mapped archaeologists will able to predict with some accuracy limiting factors for certain types of behaviour.

Heavy reliance on any one data set can leave hypotheses critically exposed. Because of the very obvious impact of environmental factors on activity in the Fens environmental data have played a leading part at every level of interpretation, at the gazetteer, parish essay and regional summary stages. Hall has summarised recent work by Waller[1] which has demonstrated inconsistencies in the description

[[1] Waller, M. 1994. The Finland Project No. 9: Flandrian Environmental Change in Fenland. East Anglian Archaeology.]

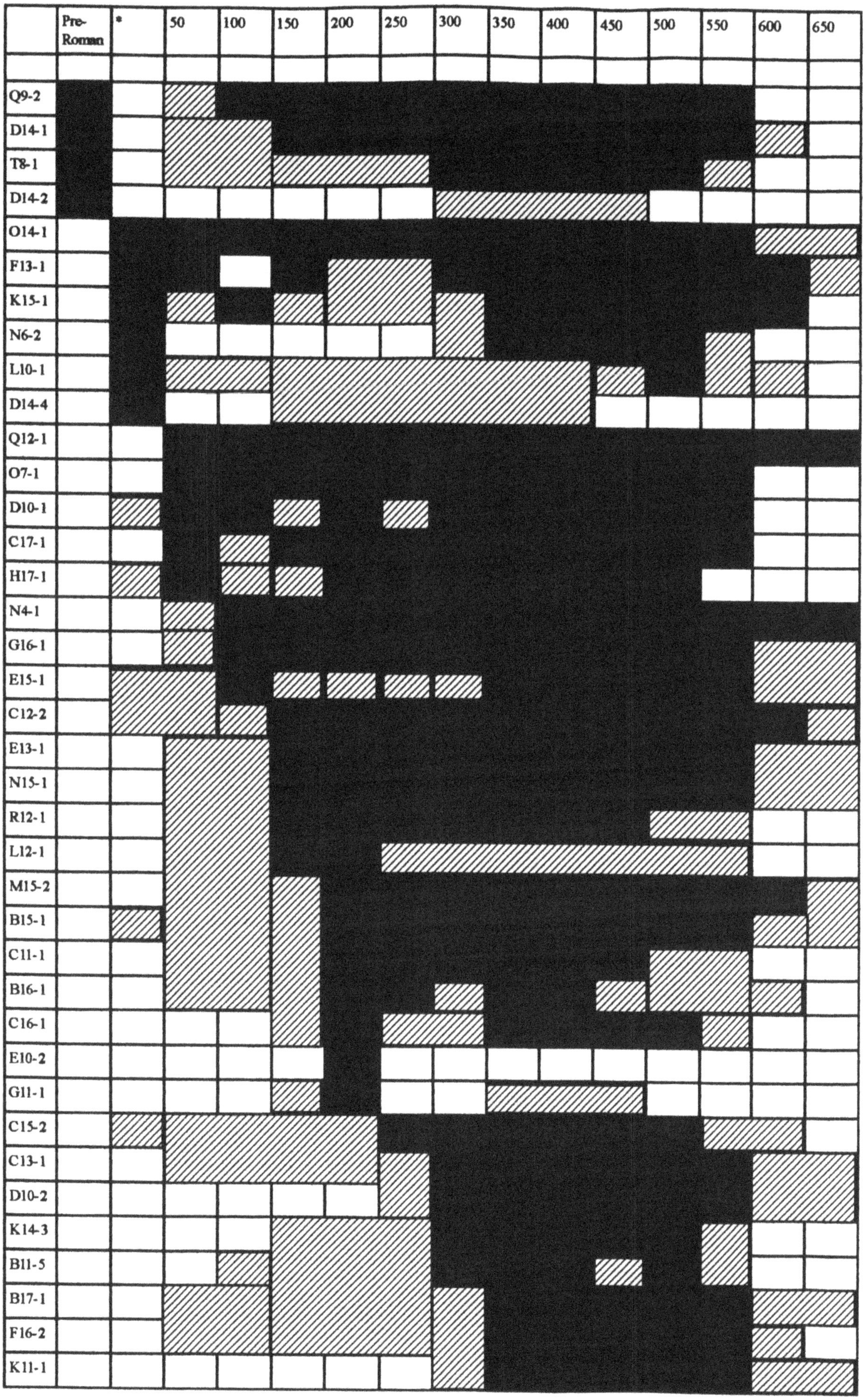

Figure 4.8. First use of sites diagram, Africa Proconsularis. The left column lists *sites* with structures. Black shading shows "statistically" more than 1 sherd, hatched less than 1, dated to a 50 year span on the site surface (from Dietz et al. 1995, 777)

of deposits, and has revised the sequence and dates of their deposition (Hall 1996, 6-8). While this should have little direct impact on positive artefactual data, it is likely to have considerable impact on negative evidence (i.e. what surfaces were exposed or created and when) since it has new implications for the loss or covering of artefacts. However, the division of the survey area into parishes, supported by "site" gazetteers does much to mitigate the effect. On the one hand, many local interpretations still stand, whilst on the other hand this presentation strategy makes it relatively easy for the old data to be reassessed in terms of Waller's findings. I have made no attempt to reassess the Lincolnshire survey examined below, because my purpose here is show how a narrative was constructed from the data as received at the time when that report was written.

For the survey, sub-regions of the fens were divided into parishes, which were subjected to desktop research prior to fieldwalking, generally conducted field by field in lines set at 30m intervals. When concentrations were encountered the intervals were reduced to 2 to 3m, in effect allowing total coverage of that area (Lane 1992, 7-8). During the process of walking soil, geology and the surface altitude were recorded, as were "recovery conditions" and crop details (Lane 1992, 9). This information was presented in a series of maps accompanying an essay for each parish. The first two maps showed the modern landscape, including its topography, and the extent and quality of fieldwalking. There then followed chronological maps showing an approximation of the landscape, along with the recognised "sites", for each phase.

The gazetteers comprised information about "sites" under nine headings: the site number, a map reference to $10m^2$, the site period, the site type, the geology, the altitude, the estimated site area, the number of pot sherds and/or flint flakes and their weights. Some entries include a few lines noting instances of burnt material or building stone, soil type and other salient features of the site. At the end of the site list was a list of find spots, referring to isolated and thin scatterings of artefacts, and below this is a list of previously known sites and findspots. Taking the example of Thurlby, in south west Lincolnshire (Lane 1992, mf E10-12), situated mainly on peat, the gazetteer shows a wide spread of Romano-British sites, mainly interpreted as settlements, but also Neolithic findspots, Bronze Age barrows, Iron Age and even Early Saxon settlements. Of the total of 29 sites recorded, all but two are on gravel. An aerial photograph of Fen-edge gravels alongside a river (Lane 1992, 157; plate VII) shows not only the extreme contrast of the pale gravels against the dark peatlands, but also highly distinctive dark cropmarks on the gravel, the continuations of which appear to be submerged beneath the peat.

There is clear evidence from "bog oaks" sealed by Bronze Age peat that during the post-glacial period, and up to the Early Bronze Age, the landscape was a woodland populated with oak, yew, beech and some conifers, but that this was drowned as water levels rose. This information was collated from Soil Survey bores. The Neolithic was represented by only a very few isolated finds, and the earliest probable sites are two circular cropmarks, which may have been ploughed out barrows (Lane 1992, 152). Plotting of ceramic data from the gazetteer or the map was interpreted as showing some Middle and Late Bronze Age settlement lying on gravels in the middle of the modern parish. Some areas of dark soil associated with increased incidence of burnt stones were also thought typical of this period (Lane 1992, 157).

By combining maps of the field collection, surface observations and aerial photography it was possible to see that "Some of the sites have been partly buried by alluvium, and pottery was found coinciding with conspicuous lines of dark occupation soil... on partly buried sites that have been disturbed by ploughing" (Lane 1992, 157). Ceramic evidence suggests that the last significant settlement before the Roman arrival was during the Middle Iron Age, with a sharp decline by the Late Iron Age. Elsewhere in the Fens this hiatus has been explained as the result of contraction or collapse of demand for salt, but Thurlby had not been an area of salt production.

The Roman period brought an increase in the frequency of settlement, some of it long-lasting. It is possible that the River Glen was canalized during this period, an argument substantiated by the durability of site 25A/25B on peat in one of the rivers old channels, implying that conditions were comparatively dry at that time. The occurrence of several "marginal sites" on the gravel fringes provides further support for the existence of a drier environment (Lane 1992, 159).

The Early Saxon period is represented by a single large site, strategically placed at the crossing of the Glen. Here, as well as bone and burnt stone, over 200 sherds of pottery were found a little over 1km from a cemetery dated to the 5th and 6th centuries. Otherwise, much of the Fen was under peat at this time, which may account for the lack of Middle Saxon pottery (Lane 1992, 161).

As for the other parishes surveyed, the interpretation of the gazetteer continued into the Medieval period. What manner of narrative does this approach produce? In the case of south west Lincolnshire the concluding remarks take us little further than the locally specific conclusions made at parish level. The authors remind us that many new sites have been found, some of them dated to periods which had been very poorly represented in the existing literature. Potential for more explicit narratives might lie in the observation that round barrows appear to have been situated at the contemporary Fen edge, probably constructed by people settled further inland. There is speculation about the transition from the Iron Age to the Roman period, specifically suggesting that the view once held that the Romans were colonising virgin land in the Fens is incorrect. It was also possible to tentatively identify two Saxon tribal areas from the disposition of finds and from the observation of natural boundaries in the landscape. This scarcely amounts to a "narrative", but the authors conceived of the survey as "a first stage in a series of complementary investigations and assessment" (Lane 1992, 256-257). They envisaged the use of targeted excavation, and hoped that their work might attract the attention of people experimenting with new remote sensing technology.

Although the Fenland project survey reports set results in wider contexts, and although they have brought to archaeologists' attention a profusion of new activity areas which have been integrated into the parish essays with information from earlier sources, its achievement has been the building of a platform from which further, more detailed work may be conducted. The environmental reconstruction is more detailed, and more effectively integrated into a wider scheme since Waller's work (see note above). Thus it can be shown that by the Middle Neolithic a clay-with-roddons phase had reached its maximum extent in the south east Cambridgeshire Fens, with a fen edge at around the 1m contour, with the exception of areas in the west where marine clays deposition was later. During the Bronze Age much of this area developed peat cover, a process which continued over the southern part of the study area during the Iron Age. Local variation is demonstrated in the north of the area by the laying down of marine silts and clays in the roddons which a few centuries later, during the Roman period, were dry enough to support settlement.

David Hall's summary of the parish interpretations for the south Cambridgeshire Fens includes small scale period distribution maps, which are complemented by Roger Palmer's aerial photographic interpretations (Palmer in Hall 1996, 192-198). The interpretations benefit from more detailed site surface and aerial data, recorded in the gazetteer, which itself includes more elaborate preliminary interpretation (Hall 1996, mf 1 and 2). This superior base level (classes B, C and D) has allowed a more specialised understanding of the significance of surface data. Thus, early prehistoric scatters of a kind which had been interpreted as industrial sites elsewhere could be re-evaluated; it is noted that they occurred where flint sources are of low quality, but within easy reach of water, quite often on light soils. Such flint would have provided a poor basis for trade, but the locations were well suited to the settlement which local barrows indicate must have existed by the Early Bronze Age (Hall 1996, 154). If the assumption that such lithic concentrations represent settlements is allowed to stand a further level of inference is possible, using negative evidence: despite the fact that the light soils provide an ideal medium for aerial photography no crop or soil marks were noted to be associated with the flints. Thus, it was suggested, that settlement structures must have been very insubstantial.

Negative evidence enabled Hall to distinguish groups of sites which seemed to represent distinct territories. The finds which did occur in the no man's land between these "subregional centres" (Hall 1996, 157), typically axes, were likely to have been employed in hunting and "exploitation of woodland resources. The importance of territorial integrity was emphasised by the apparent failure to exploit potentially productive soils in these areas. In this way class E information has been used to hypothesise the existence of a specifically class G category, the territory, which can be tested by further work focused on obtaining the kinds of data which support or undermine the proposition.

It was noted that major Early Prehistoric monuments, such as long barrows, were still emerging from the peat, suggesting that the survey's record of the total surviving archaeology of the area is far from complete. On the other hand, from the Late Bronze Age it is probable that the material record itself was deficient at a time when the pottery produced lacked durability, but when use of flint had become much less frequent. Hall suggested that various accidental finds of metalwork from that period imply a much wider settlement pattern than was visible from surface survey.

Despite increased marine incursion during the Iron Age activity in the area was more apparent for this period, with an increased density of landuse illustrated by movement onto clay islands and promontories, as well as the valleys which had been preferred in earlier prehistoric periods. The gazetteer (Hall 1996, mf 1 and 2) shows a tendency for Iron Age "settlements" to be recognised on the Roman "settlements" located by the survey. At Cottenham, of a total of 6 Iron Age sites described as "settlement", only two produced only Iron Age finds, and the visibility of each was probably enhanced by the presence of burnt stone. Four of the ten Roman "settlements" were exclusively of that period. In the survey of Eley, from eight Roman "settlements" and a hythe, four were exclusively of that period, compared with one out seven Iron Age "settlements" (Hall 1996, mf C9-10). Two other Iron Age "settlements" produced no Roman material, but of these one was within the ground of a known motte, and the other occurred in association with a Bronze Age flint scatter (Hall 1996, mf D2-5). As has been noted above there is a general tendency in northern European survey for flint and Roman pottery to survive better on the surface. Thus it is quite probable that this strong association of Iron Age sites with those of other periods has more to do a higher incidence of Romano-British and other durable material in lines at 30m intervals, resulting in the discovery of Iron Age material only during the intensive stage of work on a site. Consequently there remains the possibility that Iron Age, and Late Bronze Age sites, are significantly under-represented.

Hall believed that knowledge of the extent of Romano-British settlement within this landscape "is likely to be fairly complete" (Hall 1996, 159). While alternative methods of research would enhance the detailed understanding of this period, general economic patterns have begun to emerge. Overlapping distribution of British fine ware types was presumed to indicate overlapping trade areas. Ditch systems outlined from aerial photography suggest that large areas of the landscape were enclosed, with long ditches running from one "site" to the next, providing the tangible structure of a presumably mixed agriculture, although there is insufficient environmental evidence to prove this. Although conditions of the time would have allowed a salt industry supported by communication along a probable Roman dyke none of the artefacts found suggests one. The fabric of institutional belief of the period was represented by a shrine at Haddenham and a remarkable hoard of votive metalwork at Willingham (Hall 1996, 60, 68, 144 and 159), although neither were products of surface survey. Evidence for later settlement was presumed to have been destroyed or concealed on the sites of modern villages.

The Fenland surveys have provided an incomparable background for more intensive work, and in doing so they offer very general class F information for some periods. The political sphere is barely touched upon. A strength of the project has been the collection of geo-environmental class C data linked to ^{14}C dating. There is a substantial body of class E interpretation, but in such a large area it has not been supported by sufficient archaeological sub-surface survey for its significance to be understood. Within the gazetteers there is some class D, which conjoined with class E, information has provided tentative identification of particular activity in a specific environment (i.e. references to hunting and woodland management where sporadic flint arrowheads or axes occur), but the full value of the Fenland reports will only be realised through further work with a different emphasis.

4.4iic The Biferno Valley, Molise

It is interesting to make a crude comparison of the proportion of space devoted to different components of reports. Hall devoted some seven pages to his overall narrative summary of the Isle of Eley survey (Hall 1996, 154-161); Gaffney and Tingle devoted 15 pages to the reconstruction of the Maddle Farm villa economy alone (Gaffney & Tingle 1989, 224-238). Dietz summarized two large volumes in 27 pages, many of them taken up with maps (Dietz et al. 1995a, 772-799), while Barker lavishes 170 pages on the human activity in the Biferno Valley upto AD500, and a further 53 from then until the later 20th century (Barker 1995, 84-253; and 254-307). His narrative is supported by detailed data presented in the companion volume (Barker 1995a). Each defined period received interpretative treatment under the headings "settlement trends", "settlement structure", "environment and subsistence" and "material culture", which were combined to render a social-economic account, including matters of belief and political or hierarchical structure.

Surface collection was conducted at three levels. The narrowest interval between walkers, 15 metres, was described as intensive, and there were also intermediate and reconnaissance levels, the intervals for which have not been specified in the report (Barker 1995, 43-46; fig. 21). The methods by which geomorphological information was obtained have been outlined above (*3.3vib*).

From the outset Barker was concerned to establish good reference collections from properly excavated stratified deposits for all periods. The Early Neolithic artefact chronology depended largely on material from a site located during survey, and subsequently excavated, at Monte Maulo (Barker 1995a, 18; Barker 1995, 100, 104). Some 1500 Early Neolithic sherds of various types, many of them impressed ware, were found in a stratified deposit and pit fills sealed by it. The deposit was in the process of being eroded by plough. Four ^{14}C assays dates were consistent with the pottery, providing a good basis for dating co-occurring lithic material. The lithics were less prolific, comprising 58 flakes, 3 of which were retouched, 6 blades, 19 cores and 10 waste flakes and other fragments totalling over 200 pieces, all but five of which were of poor quality chert. The characteristic features of this assemblage were used to designate other surface assemblages as "probably neolithic". A Middle and Late Neolithic group was identified by flake characteristics (in particular, parallel-sided blades) which have far wider currency (Barker et al. 1995a, 100) and, where ceramics occurred, by finewares of that period.

The distribution of Early Neolithic finds (Barker 1995, 101; fig. 39) reveals a cluster of concentrations, mostly on the south east side of the river in the middle (A336, ?A337, ?B282, A314-316; these on promontories) and lower parts of the valley (B198 (Monte Maulo), A268, C186, ?A276), and one site (A184) at the coastal end of the valley in the hills above the north west side of the river (Barker 1995a, 12-25). These are variously described as domestic, stone-working and sporadic sites. Three of them, all of which produced material from Early to Late Neolithic phases, were subjected to test pitting (A268, C186 and A184; see Table 4.4).

Whilst this list shows the potential for discrimination between sites, the data must be treated with much caution, as they are not comparable. Finds recovered during excavation and test pitting have not been distinguished from those recovered on the surface. It is not clear from the gazetteer data what distinguishes a *domestic* from a *?domestic* site. From the site class it is possible to get an impression of the frequency of period specific artefacts, but the variation in numbers of sherds and flints is more a reflection of the recovery technique than the density of surface scatter. Only recourse to the archive would show the sum of material judged to be Early Neolithic.

There appears to be a distinct pattern of topographical preference, but the sites occurred on precisely those features in the landscape where finds are most likely to be exposed on the surface, in areas where heavy erosion is known to have taken place. As has been pointed out above, the preferred settlement locations for whole periods have been shown to be concealed beneath colluvial and alluvial deposits, and it is clear that there has been substantial soil movement during several phases since the Neolithic in the Biferno Valley.

Environmental evidence in the form of bone was recovered from the excavation at Monte Maulo, and from the test pits at C268, A184 (from which ^{14}C dates were obtained) and C186 (where dating was by association with pottery; Barker 1995a, 9-26). Plant remains were recovered from Monte Maulo. The latter site alone provided secure Early Neolithic environmental data, and only a small sample. The minimum number of individual animals securely represented was only one from each of three species: cattle, sheep/goat and pig (Barker 1995a, 144). These animals had reached maturity at the time of death suggesting, in the first two cases, that they had been maintained for their secondary products as well as meat. The small carbonised grain assemblage was dominated by the cereals barley (38 seeds) and emmer wheat (78), and legumes included lentil and pea (it is unclear why the text states that barley "dominates" wheat; the statement appears to be at odds with the trend noted in Barker 1995, 111). The molluscan evidence implied a locality comprising open and covered ground. Barker acknowledged that although the evidence demonstrated the existence of an agricultural economy, it could not be used to show the balance between meat and plant production.

The summary narrative, under the heading "Neolithic Societies in the Biferno Valley", followed presentation and discussion of the material record and was set within a framework of existing literature, from which it was assumed that Neolithic "societies developed systems of exchange to maintain and promote social cohesion" (Barker 1995, 127) which might have proceeded from "risk-avoidance" strategies. Thin-section analysis showed that the fabrics of the finest wares were from elsewhere, but that most fine and coarse wares were composed of local fabrics. Nonetheless, the importance of exchange was emphasised by the presence of non-indigenous high-quality lithics which indicated "considerable interaction between neolithic groups throughout the peninsula" (Barker 1995, 128).

In looking at ritual during the Early Neolithic Barker was compelled to cite examples from Molise's northern and

Site	Location	Topography	Pot	(count)	Lithic	(count)	Class*	Interpretation
A314-316	mid vally	spur	coarse	40	cores/flakes	much	5c	work site
A336	mid valley	spur	coarse/dec	11	cores+flakes	4	4b	?domestic
A337	mid valley	spur	dec	1	none	0		?domestic
B282	mid valley	spur	dec	?	flakes	118	4c	domestic
B198	low valley	slope	coarse/fine	1500	cores/flakes	68	4c	domestic
A268	low valley	terrace	dec/medium	100+	cores/flakes	64	6c	domestic
A269	low valley	terrace	dec/?	5	flakes/blades	5		domestic
C186	low valley	hilltop	dec/?coarse	3+	?cores/flakes	?	6c	domestic
A276	low valley	saddle	dec	1	flake	1	6a	?domestic
A184	above low	hilltop	fine	?much	cores/flakes	31	4c	domestic

[NOTE: dec - decorated. Class 4 represents a "surface scatter or concentration" of upto 50 x 50m; class 5 is greater than 50 x 50m, upto 100 x 100m; and class 6 is larger than 100 x 100m. "a" represents a density of less than 1 artefact per sqm; "b" is 1-2 artefacts and "c" is 3 or more artefacts per sqm. These classifications reflect overall artefact density, and in most of the above include material from more than one period.]

Table 4.4. Summary of Neolithic findspots in the Biferno Valley southern neighbouring regions, Abruzzo and Apulia[1]. The survey found no direct evidence for this area of behaviour, and if the objective of Barker's book was merely to provide an interpretation on the basis of the survey's fieldwork introduction of this level of narrative would be inappropriate. However, so long as the source is fully explicated, the wary reader should have no qualms about having the opportunity to consider this as an untested model or hypothesis which amplifies our view of potential human activity in the region.

The qualitative data from the survey of the valley was not sufficient to build a hypothesis although, at the very least, it did not contradict the hypotheses promoted for neighbouring regions. But if the site, as a determining unit of survey, is so thinly represented in terms of distribution over the region, it cannot be used to support general assertions about population density, and is of limited application in discussions of interaction and exchange, unless the rarity of occurrence is itself demonstrably representative of actual spatial patterning. The geomorphological survey (Barker 1995a, 59-82), coupled with the proportional variability in extent of surface coverage within the valley, makes such an assumption unsafe and, in the present state of archaeological technology, untestable. The justification for this narrative freedom lies in the author's situational intent in temporal and spatial dimensions at every stage of the story's evolution; successive phases emerge from earlier phases, and from neighbouring regions.

A sterner test of the relationship between sampling, analysis and narrative comes from the analysis of a period where data are more prolific and for that purpose I have chosen the Iron Age, which has the advantage of not having been subsumed into classical accounts, so that the archaeology can speak louder than the ancient texts. The criteria for diagnosing sites were as for all periods after the Neolithic: more than 20 sherds were considered to be probable sites, 5-20 were considered possible and less than 5 were described as "sporadic" (Barker 1995, 160). Daunian and *impasto* wares provided fairly well understood ceramic forms and fabrics. 114 recognised findspots of the period broke down as 58 probable and possible sites and 12 findspots in the lower valley, 30 and 1 in the middle valley and 12 and 1 in the upper valley, with no findspots of any kind in the Matese mountains. Some 56 of the best established sites were considered to be domestic. The bulk of sites presented as surface scatters covering an area of less than 50 x 50m, but indications of the emergence of a very distinctive site hierarchy were represented by a few sites which covered up to ten times greater areas (Barker 1995, 161-162). It was noted that the total number of findspots exceeded that for the Bronze Age by nearly 50, and given that the period designated as Iron Age was of 500 compared to 1000 years there appears to have been a sharp increase in the amount of settlement in the valley.

The most detailed evidence of settlement structure came from a site at Arcora, on the coast, which was located by amateurs and was excavated subsequently under the auspices of the state (Barker 1995, 163-164). Finds demonstrated activity on the site from the Late Bronze Age, but well built sub-rectangular structures were associated with two major phases in the eighth to seventh and sixth to fifth centuries BC, when the settlement appeared to have developed into a village. Some ceramics illustrated the continuity of Bronze Age traditions, but of particular note was the frequent occurrence of spindlewhorls and loomweights, strongly implying the importance of textile production to the community.

The survey located three significant concentrations of material on a quarried spur to the north of the river at Santa Margherita, in the lower valley, within approximately 700m of each other (Barker 1995, 164; Barker 1995a, 37; map). These sites, A90-92, probably formed discrete components of one large settlement, heavily disturbed by the quarrying, which offered section views of pits during the survey. A90 was regarded as a high concentration over a medium sized area, including 55 Daunian and 10 *impasto* sherds. A91 comprised a small, thinly dispersed, scatter of 4 impasto sherds, and A92 a moderate scatter over a larger area of 18 Daunian and 25 *impasto* sherds. The whole was interpreted as a village which continued into the Samnite and Roman periods (Barker 1995a, 7).

Other areas of large scale Iron Age occupation were diagnosed at: B287, a mid valley ridge some 5km south of the Biferno, where 1160 sherds of an undecorated ware were considered to be a Daunian type, and where 67 *impasto* sherds were also recovered from the surface. Subsequently this site developed into a major Samnite hillfort (Barker 1995a, 20-21). At E11, a lower mid valley ridge 5km east of B287, 28 Daunian and over 100 *impasto* sherds represented a large concentration (Barker 1995a, 33), and at Monte Vairano (D32), a hilltop in the upper valley, was also assigned to the Iron Age, although diagnosis could not possibly have been made on the basis of surface finds from the survey presented in the gazetteer (Barker 1995a, 31), since none of the material appropriate to the period is listed. No other corroboration is offered since Barker fails to mention any Iron Age finds from the small scale excavations carried out at this 50ha Samnite hillfort under De Benedittis (Barker 1995, 188; Hayes 1985).

A cluster of medium density Daunian/*impasto* scatters occurred amongst classical material in the lower valley on ridges (A202/A203/A206), and on a spur (D101); and of *impasto* only on a ridge and spur in the middle valley (A142/A146), over a small area on a slope in the lower mid valley (B237). These locations are variously interpreted as small domestic, farm or cemetery sites (Barker 1995a, 8-32). Other scatters include material with Late Bronze Age characteristics, which might be of Iron Age date (for example, C284/C285).

Although Daunian and *impasto* sherds co-occur in the middle and lower valley it was noted that Daunian wares did not occur in the upper valley. This was taken to reflect channels of communication which operated in an Apennine and a coastal zone. Although isolated items of Greek style statuary have been found at the upper end of the valley there was much more material evidence for links with Campania. A different picture emerged from the lower valley, where the Molise showed more affiliation with an Adriatic coastal zone which included Abruzzo and Apulia, an area upon which Greek culture had an impact in the late Iron Age. These cultural distinctions appeared to be more pronounced in the

---

[[1]Skeates, R, 1991: *Caves, cult and children in neolithic Abruzzo, central Italy* in Garwood, P; Jennings, D; Skeates, R; Toms, J (eds.): Sacred and Profane, pp. 50-64. Oxford. Whitehouse, R, 1992: *Underground Religion: Cult and Culture in Prehistoric Italy.* London.]

following Samnite phase when sub groups named in classical literature can be placed at different ends of the valley.

The Iron Age settlement hierarchy was perceived to range from quite densely settled villages, to small villages and hamlets or farmsteads akin to those of the Bronze Age. The persisting importance of emmer wheat and barley as staples is suggested by plant remains from Santa Margherita and Arcora; pea and horse bean are also represented, the latter dramatically by over 30,000 beans from a "store" at Arcora. Unfortunately, this discovery may say more about the problems of representing a pattern of production on the basis of the otherwise very small, localised samples from just two sites. Carbonised plum and grape were also found, and Barker treats the occurrence of fine Daunian drinking vessels as indirect evidence for the latter being a cultivated variety, supplying a local elite (Barker 1995, 168-170).

The faunal data for this period is very limited. The bone from excavations at Arcora has yet to be analysed, and only 458 fragments of an already small sample from Santa Margherita were from Iron Age contexts. Again, similarities with the Bronze Age record emerged, with sheep/goats predominant over cattle and pig, although the presence of horse and donkey (Barker 1995a, 145) gave Barker grounds for speculating about these species potential significance, one as the vehicle of the elite, the other as a beast of burden, with a particular niche in the haulage sector of the economy.

More cogent evidence for agricultural intensification was derived from a second level of analysis. Survey and excavation data, particularly from Arcora, suggest a trend towards agglomeration of the population, coupled with high instances of craft tools such as the spindlewhorls and loomweights required for textile production. This implied both a degree of centralisation, and of specialisation of production (Barker 1995, 171). However, such a hypothesis could only be tested against good quality, probably excavated, data from a wider variety of sites.

Cemetery data from lower valley excavations at Porticone (near the coast), Santa Margherita (69 and 9 excavated graves, respectively) and Larino provided sizable grave samples, but in the first two cases any assumption should be tempered by the knowledge that they span the late Iron Age and the early Samnite periods.

All three of these cemeteries showed marked gender differentiation. Most graves included Daunian and/or *impasto* ceramics, but females were usually associated with jewellery, and men with weaponry. The Santa Margherita grave goods were richer, some including bronze vessels, emphasising the hierarchical differences between settlements. Five of these latter graves were demonstrably 6th century BC, but the others, dated to the 4th century, were from the Samnite period, suggesting either a break in the use of the site, or that there may have been other graves. A break would undermine assumptions of continuity of tradition. There was no detailed chronological distribution given for the graves at Larino and Arcora (Barker 1995, 171-172). The lower middle and lower valley cemeteries had in common the presence of Daunian and *impasto* wares. The cemetery at Campochiaro served to bolster the argument that there was a distinctly Apennine flavour in the upper valley. The graves there included *impasto*, but in place of the Daunian material were finewares common to Campanian cemeteries.

Hierarchy manifested in settlement, and in distinctions between graves was more marked beyond the valley. In Abruzzo the inclusion of high status goods provided evidence for the importance of lineage over age (Barker 1995, 177). Here, and in other neighbouring regions, the evidence is for a chieftain-led society, in which males assumed the warrior mantle. The survey data suggest that the people of the Biferno Valley were slower to adopt such traditions, which only became a full part of their lives in the ensuing Samnite period. Nonetheless, surface and excavated finds from within the study area are enough to suggest that there was a trend towards more elaborated hierarchy.

In looking at the presentation of narrative for the Early Neolithic and Iron Age periods in the Biferno I have deliberately made light reference to Barker's use of information from outside the valley. His concern is not merely with narrative, but rather with synthesis, so that the breadth and depth of his acquaintance with the literature should not be seen as a means for concealing the lacunae in data from the valley itself, but rather as part of the wider spatio-temporal flux within which the valley's history is situated. Nonetheless, there are moments when a lack of finds within the valley is obscured by extrinsic reference, and for that reason I have tried to focus is on data recovered during the survey, coupled with Barker's integrated accounts of Superintendency excavations.

Barker himself has been critical of the field methods. Sampling in the middle valley was inadequate, and walking intervals appear to have been inconsistently applied. The failure to record collected data at a fixed area ratio makes a full evaluation of the gazetteer record itself impossible, and must surely have presented Barker and his colleagues with problems during interpretation when they seem to have relied on estimated rates of surface density. It would be impossible to make wider statistical inferences about population distribution, or even to estimate reliability. Equally problematic was the classifying of the site densities and their extent since definition was based on material from all periods; only the site function interpretation attempted to be period-specific.

These faults are in the main products of their generation. In other respects the survey was ahead of its time. The recognition of the need for better understanding of ceramic sequences and lithics at regional level has only recently become a general concern in Mediterranean surveys. Although the excavation and test pitting strategies employed by the survey were neither systematic, frequent nor productive enough to stand rigorous examination they provided environmental and material culture data which combined to offer glimpses of human behaviour which, when set against the background of Barker's review of other work in the area, can be tested by further work in the field. No less important was Barker's use of Chris Hunt's geomorphological work (Barker 1995a, 59-82) to illustrate the radical changes that occurred in a landscape during brief periods of geological time and the use of geophysical prospection within a survey was certainly a rarity at the time.

I have focused on two relatively unproductive periods in an effort to avoid the intrusion of historical narrative upon archaeological interpretation. Consequently I have omitted discussion of Barker's interpretation of belief systems, based on survey data which were elaborated very effectively for the Samnite period, as were the changes in the material record of the Roman period which suggest new ideological allegiances. In doing this I have deliberately highlighted problems within an overall narrative with which, from within the Mediterranean, only the Southern Argolid report can compare (Jameson et al. 1994). There the field methods were better controlled, but the presentation of data is insufficient, and the integration of narrative is less effective. I suspect that upto the present Barker and, in later chapters, John Lloyd, have constructed the most effective and compelling class G accounts of proto-urban and urban cultures as they emerged through second half of the 1st millennium BC and flourished and declined in the subsequent centuries.

4.4iid Danebury Environs Programme

It would be easy to be dismissive of a programme which includes the word "environs" in its title, but then limits fieldwork to geophysical survey and excavation in just seven small areas: it appears to flout every possible rule of sampling. At the time when the project was being formulated the Wessex chalk of central southern Britain could probably claim to have received more archaeological attention than any other rural area in Britain. In addition, Rog Palmer had recently published his air photographic interpretation of the area (Palmer 1984), whilst Barry Cunliffe had just completed twenty seasons during which he had excavated 57% of the interior of Danebury hillfort. The right programme might indeed allow an "entirely new understanding of social and economic organisation... from *c*. 1400 BC-AD 100" in that landscape. The conventional approach would have been to fieldwalk land under cultivation within a sampling strategy, but for reasons unspecified surface collection played no part in the Programme (some had taken place during the early excavation seasons at Danebury. Cunliffe 2000, 9).

A reason for avoiding fieldwalking may have been its technical unreliability, although at the time the project was designed there was little published about the problem. It seems more likely that Cunliffe wanted to avoid basing analysis on formless pieces of prehistoric ceramic fabric, and preferred instead to look for better quality, reliable artefacts from excavation. He also had a rich resource from previous excavations within the study area (Cunliffe 2000, 34-37) which included: Middle to Late Bronze Age enclosed and unenclosed settlements, cemeteries and linear earthworks; and Iron Age hillforts, enclosed and unenclosed settlements.

Areas for fieldwork were selected from the air photographic evidence on the basis of the information they might provide about the wider organisation of the landscape, where they were expected to fall within a period spanning the Late Bronze Age and Iron Age. As a first stage areas of usually several hectares were surveyed by magnetometer showing much of the structure of the targeted area, including some finer detail such as hearths and pits, and sometimes tying the targeted area into wider landscape features. The excavations were then carried out to the standard set at Danebury, including a long linear, a field system and a variety of enclosure types. With the exception of the total excavation of the Banjo enclosure at Nettlebank Copse all the excavations sample only parts of parts the landscape features under investigation. The assembled data was of a quality and quantity far greater than would be acquired from a surface collection, including ecofacts from securely phased contexts. Consequently it was possible to build a much richer picture of events within a particular space, often with implications from beyond that space (ecofacts and artefacts or raw materials which came from further afield).

Whilst the results generated an interesting narrative there must be doubts about the appropriateness of making generalisation about the wider landscape, in particular assertions based on negative evidence. There are many features that do not show up on air photographs, notably stake and post holes, hearths etc., which represent the most likely targets for an unenclosed settlement. Since the geophysical survey was largely focal (with the exception of an extensive length of a long linear cutting the Windy Dido field system. Cunliffe & Poole 2000g. Figs. 7.1 to 7.3) it is difficult to see how hypotheses can be asserted regarding the general extent or density of settlement. It is not sufficient to argue that the earlier work carried out in the is large survey area would have picked up small, settlements, or even that watching briefs over recent developments would have identified material of a particular period if it was there. After all, the earlier excavations were themselves targeted at visible landscape features.

However, the detailed ceramic sequence allows a far greater level of confidence in the phasing of activity in areas excavated within the study area, either as part of the project, or previously. It was possible to show internal site developed, and to make comparisons with other sites which were contemporary. With contemporaneity established is was possible to consider variations in function and status between sites. Whilst the programme has been hugely valuable for British Iron Age studies its status as an environs programme remains unsatisfactory.

4.4iie South Cadbury Environs Project

The project began very informally as a judgement-based programme of fieldwalking attempting to provide evidence contemporary with activity on Cadbury Castle, an Iron Age hillfort whose earlier excavations were being reported by a team from Glasgow University (Barrett et al. 2000). In 1992 Paul Johnson of Glasgow introduced the use of gradiometry and resistivity and by 1996 a clear research plan had evolved in which geophysical survey played the leading role. In the first place the extent of the study area had to be defined. The surveys around Maiden Castle and Danebury provided models for different techniques and extremes of scale for fieldwork in hillfort hinterlands. In contrast to the other two there had been very little air coverage of the Cadbury area, paradoxically because of its situation under the flight path of a military airbase. Without the potential for the construction of hypothetical phased landscapes for testing over a wide area it was decided to limit the study area to an 8 by 8 km square, similar to that at Maiden Castle.

The study area is centred on Cadbury Castle, Somerset (Tabor 2002, 6-7). To the southwest, extending to the northeast the hillfort is bounded by an arcing ridge and plateau of Upper Lias sands capped with Inferior Oolitic Limestone, from which it is an outlier, separated by vallies to the south and

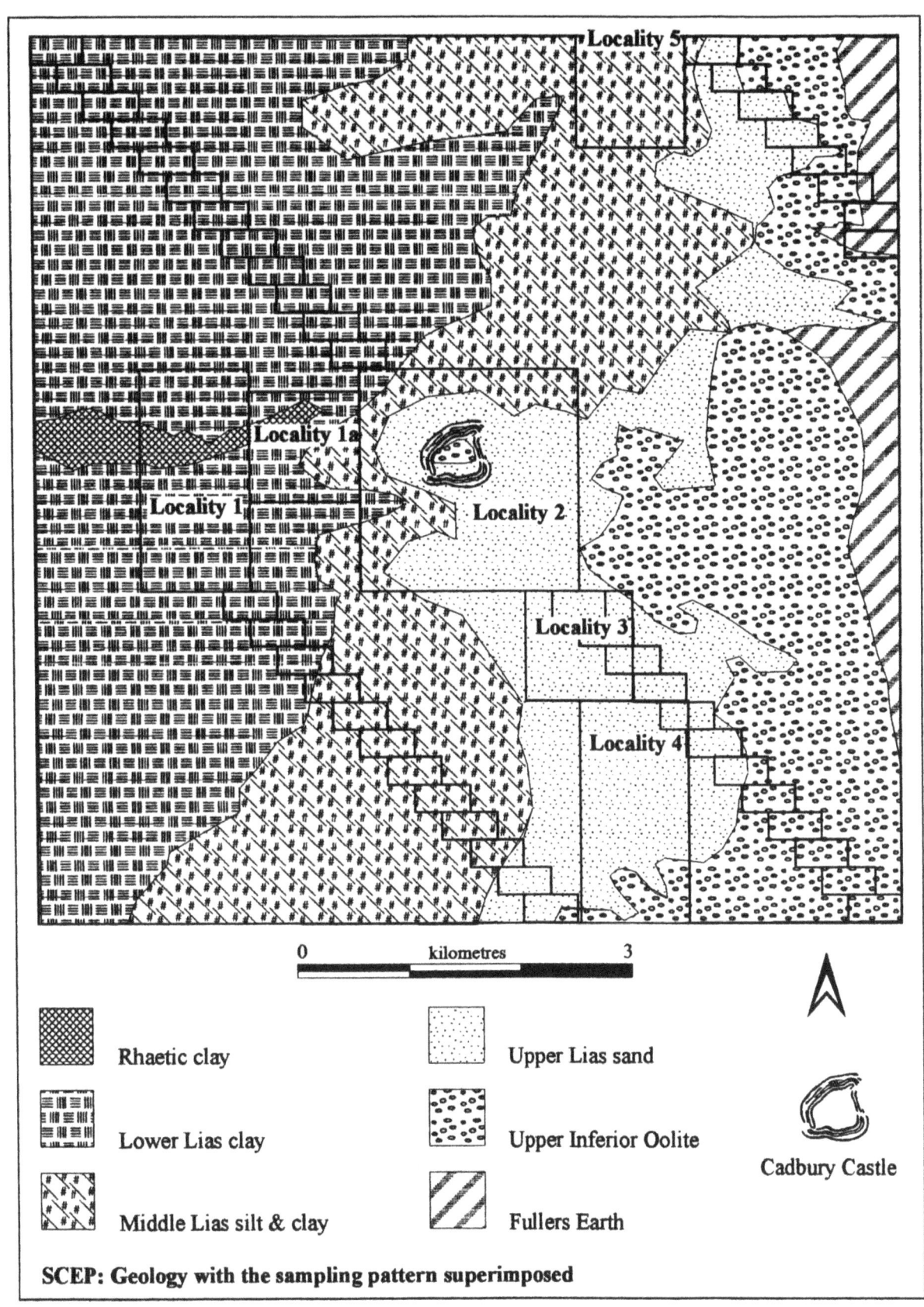

Figure 4.9: Geology with the study area, with sampling areas superimposed, South Cadbury Environs Project

east. To the north the near horizon is formed by a ridge of Middle Lias silt and clay, but there is an open aspect to the lower lying and generally flatter upper lias clays, with some limestone outcrops, extending from and beyond the north west and west of the study area (Figure 4.9). In general the land is given over to sheep and cattle, although there where cultivation has taken place it is clear that the archaeological resource is being damaged. Maize is an increasingly popular fodder crop and potatoes, until recently confined to the lower ground, have now been introduced to the plateau.

When the project was designed it was suspected that the soil conditions and depths would vary greatly from the high ground to the low ground. This might have impact on both the efficacy of geophysical survey and the recovery rates in surface collection. There was also concern that the limited and seasonal availability of ploughed land would bias a programme of fieldwalking, hence shovel pitting was selected as for ploughzone sampling. It was presumed to have the added benefit of reducing performance variation to due light and ground conditions and differing perceptual skills. Approximately 60 litres of soil were collected and sieved through a 1 cm mesh at all the corners of grids set up for geophysical survey, oriented to the Ordnance survey. 1 x 1 metre test pits to natural or a maximum of 2m were dug in every hectare, sited at every round 100m on an adapted Ordnance Survey 1:2500 map, to show the relationship of surface data to stratified material and to gain geomorphological information, which both informed the narrative and enabled evaluation of the geophysical data. Where the depth is considered too deep for there to be a reliable correspondence between top or ploughsoil and the strata it sealed additional test pits are excavated, usually targeting geophysical anomalies.

The full planned programme is two-tiered. Present work is within contiguous rectangular localities of 1 to 4 sq km representing the range of geology, topography and drainage. Future work will be in four contiguously segmented transects bisecting the topographic grain (Figure 4.9). The intensive techniques prevented total coverage of the targeted areas so it was decided that each should be subjected to 20% coverage. A pilot study at Sigwells (locality 3) showed that large survey areas were the most likely to generate comprehensible survey data and so it was decided to work in large blocks rather than with more frequent but smaller units (Tabor & Johnson 2000).

The project's principal target period spans the Late Bronze Age through to the 1st century AD, although data from all prehistoric and Romano-British phases are fully recorded, and the combination of test pitting and geophysical survey has proved particularly effective for this. Ploughzone sampling has been less so, a reflection of soil depth in the valley, and of the hostile environment for friable pottery on the plateau. There are also problems with accurate phasing of the pottery. At the project's inception there existed a comparative database from excavations on Cadbury Castle, at the centre of the study area, which had produced an extremely large assemblage, thought to comprise a range of fabrics which varied in a distinctive sequence through the whole of the 1st millennium BC and into the 1st century AD (Alcock 1980). It had been based on material from inner rampart sections, particularly one revisited in 1973, three years after the main programme had been completed. There was no strong link to an absolute chronology with the best established radiocarbon dates beginning to the beginning and end of the period. Ann Woodward found it difficult to replicate the sequence in the interior (Barrett et al. 2000, 28-43). The situation has been obfuscated still further through the discovery during the course of the project's excavation's that a common fabric considered to originate at around 500BC was already current four centuries earlier, whilst the most characteristic Middle Iron Age fabric can offer specimens of form clearly attributable to both the Early and Late Iron Age. Clearly this is a problem where a landscape survey relies on ploughzone sherds which are often little more than fabric without form.

The test pits and training excavations have mitigated the problem through the recovery of formally diagnostic sherds. However, the persistence of indigenous forms and fabrics into the Early Romano-British period make it difficult to distinguish between pre and post conquest activity. These are crucial issues for a narrative which aspires to address the character of settlement before and after the emergence of the developed hillfort, and perhaps even more so when attempting to diagnose the character of the Roman occupation from AD 43-70, and the response of the occupied people. Where individual test pit contexts produce a relative abundance of material with a multiphase currency it is probably sufficient to use the absence of common types which overlap with their later phases as a strong indicator of an earlier date, but the small samples recovered from most test and shovel pits are poorly suited to such a procedure.

Figures 4.10 and 4.11 plot all Middle and Late Iron Age pottery recovered from test and shovel pits from Localities 1, 1a, 2 and 3. Whilst it is very clear that during the Middle Iron Age there is a focus of activity around the hillfort (Locality 2) the TPs have been less effective in demonstrating activity on the hilltops to the east and south east. Immediately to the east, where the plateau has been severely degraded by the plough, some 200 Middle Iron Age sherds were recovered in a buried soil sealed by headland plough soil at the top of a scarp. At Sigwells, 2 km south east of the hillfort, three of twelve trenches have produced Middle Iron Age pottery, some from an enclosure ditch overlooking the hillfort. What the programme has revealed is strong evidence for settlement around the hillfort at a time when it has been considered absent around other hillforts at comparable stages in their development (Danebury in *2.5v* and Maiden Castle in southern Britain). It may well be that the choices of research strategy are more responsible for the negative evidence than any lack of prehistoric residues. But SCEP is not yet able to compete with the inter and intra site chronology of the Danebury Environs Programme, a flaw which will only be rectified with more excavation linked to a programme of absolute dating. Some of the pottery plotted for the later period is of ambiguous determination, and might as easily be attributed to the Middle Iron Age (in particular, the two largest groups immediately north and two kilometres west south west of Cadbury Castle). However, the Late Iron Age assemblage ought probably to be substantially larger than has been represented here as much of the material assigned to the Early Romano-British period seems increasingly likely to be very Late Iron Age.

Where SCEP gains over other projects by being able to link successive landscape architectures to the distribution of good

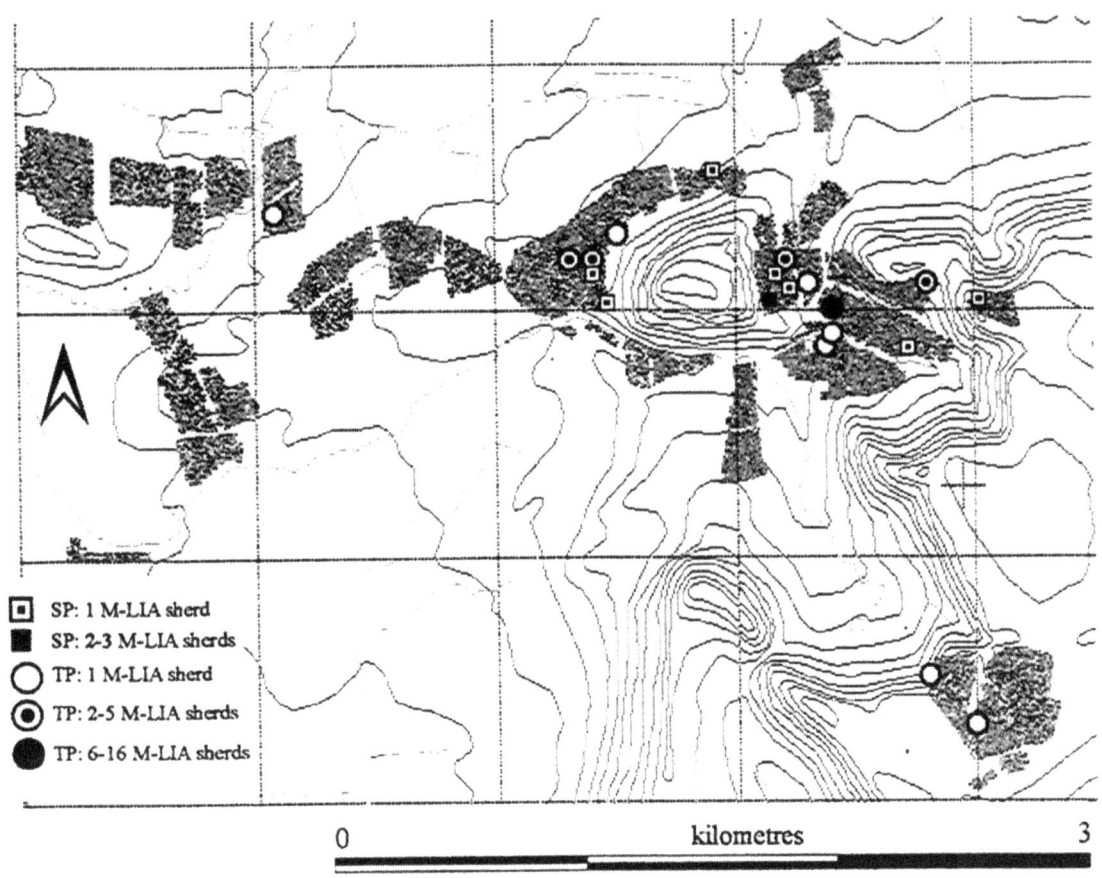

Figure 4.10: Distribution of Middle to Late Iron Age pottery superimposed on gradiometer plots, South Cadbury Environs Project

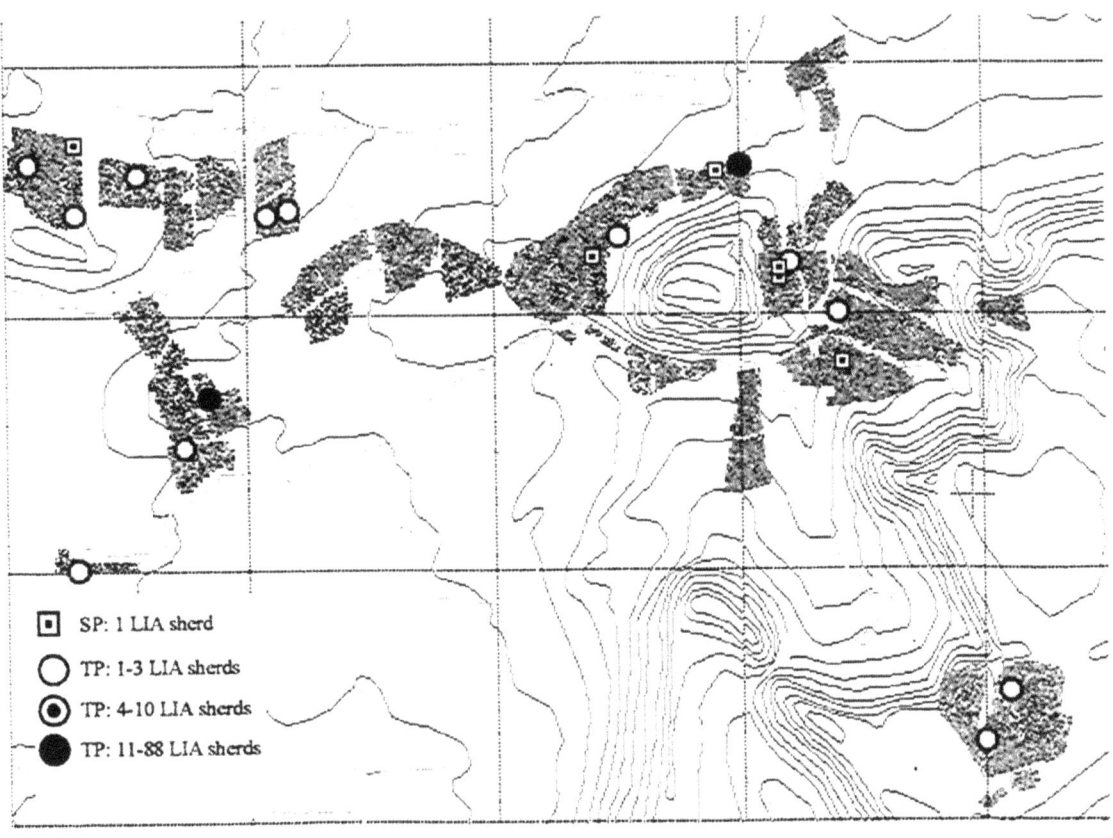

Figure 4.11: Distribution of Late Iron Age pottery superimposed on gradiometer plots, South Cadbury Environs Project

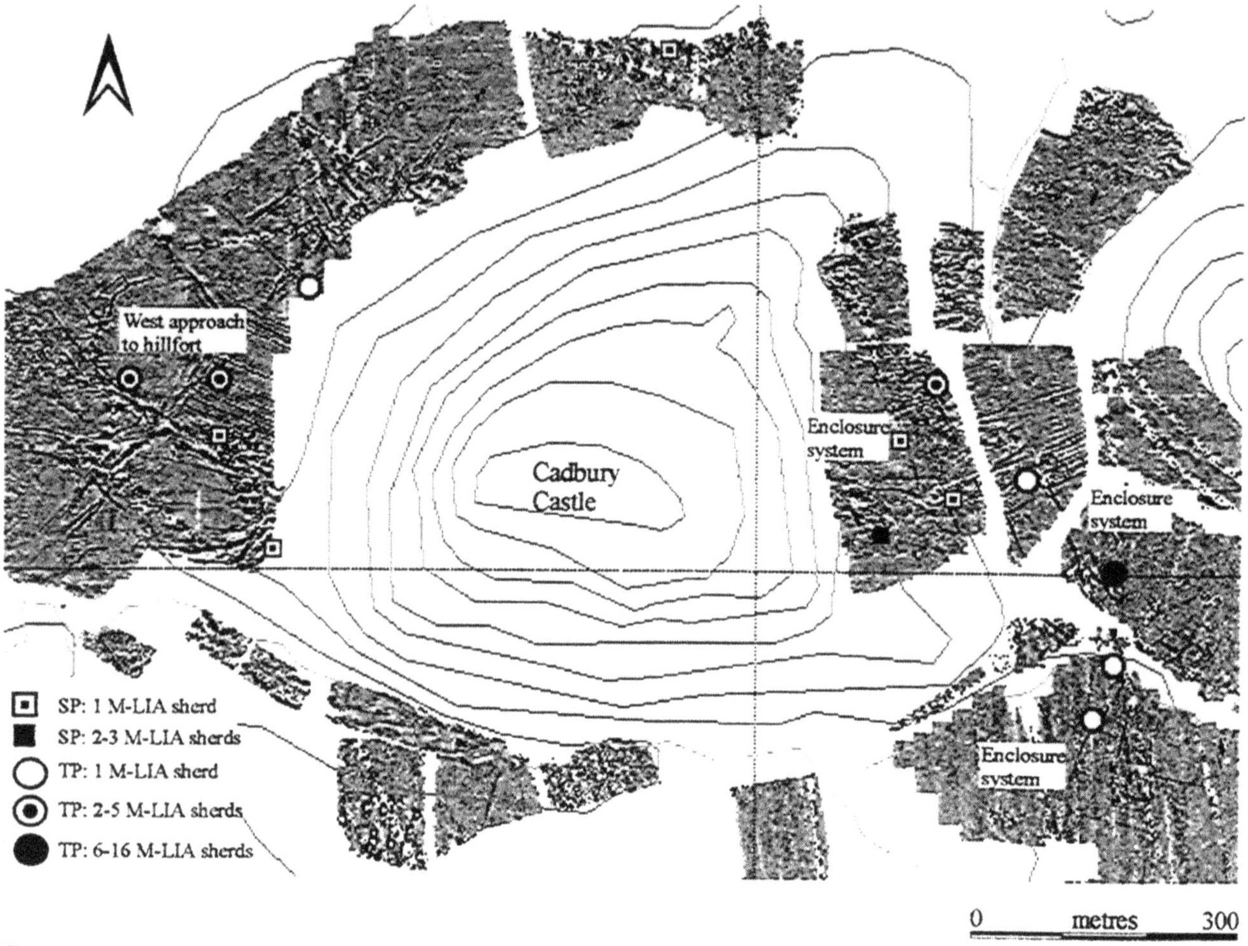

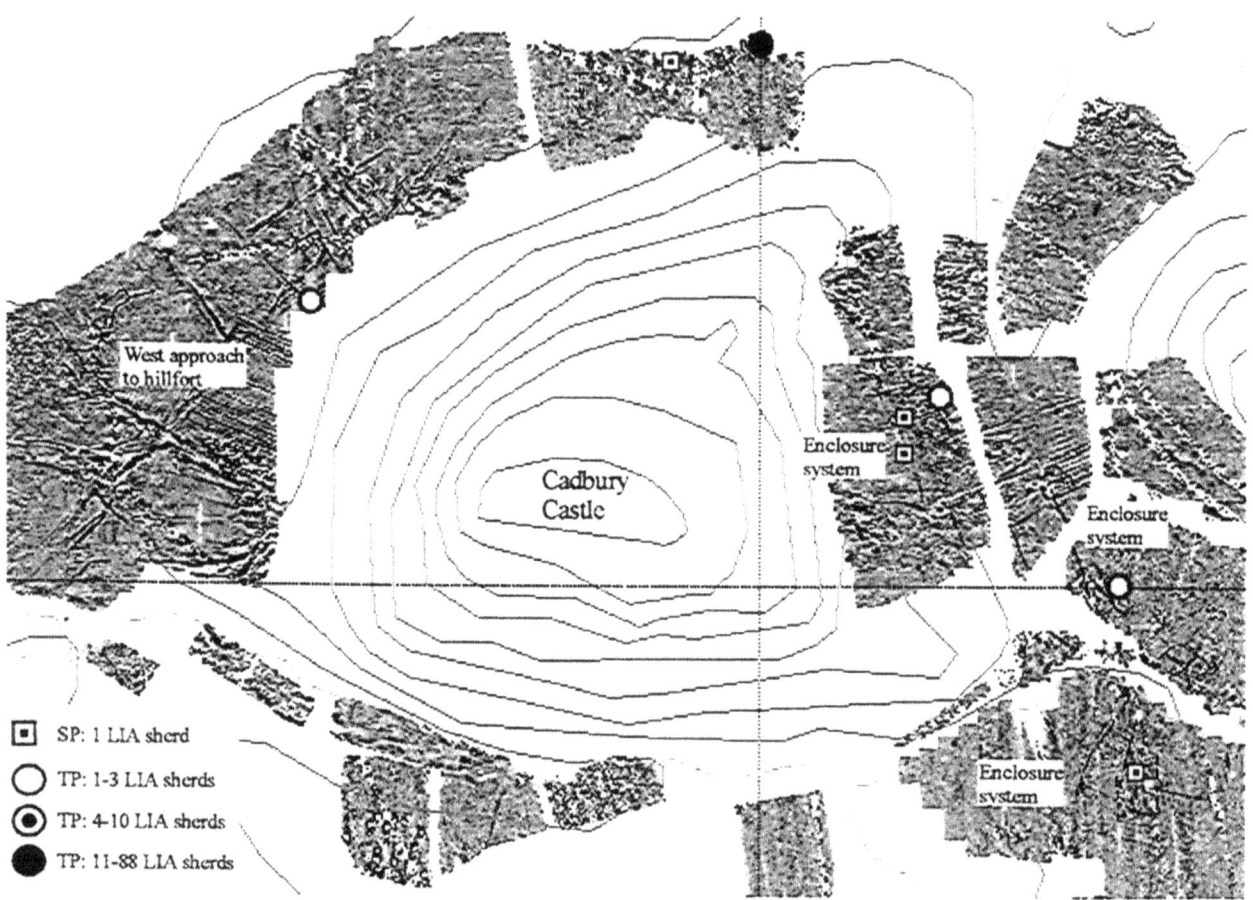

Figure 4.12: Detail of distribution of Middle to Late Iron Age pottery superimposed on gradiometer plots, South Cadbury Environs Project

Figure 4.13: Detail of distribution of Late Iron Age pottery superimposed on gradiometer plots, South Cadbury Environs Project quality sherds. Figures 4.12 and 4.13 shows details at a scale large enough to reveal enclosure ditches which appear to be associated with specific ranges of finds. The former shows how a ditch-defined funnel towards the south west gate should probably be regarded as integral to the developed hillfort, but that it was no longer a focus of activity by the Late Iron Age. In contrast enclosure systems to the east and south east of the hill appear to have remained in use into the first half of the 1st century AD at least. It is the linkage between finds and geophysical data which gives interpretation its narrative thrust.

4.4iii   Seeing believing: control and accountability

The political and religious beliefs of pre-literate societies existed as mental property which was recreated by, and found expression in, the customs of authority and rite. Before presuming to describe belief we must be able to reconstruct the activities associated with it, and to interpret the *significance* those activities had for their original participants. This is a less straightforward task than, for instance, observing the distribution of millstones with respect to carbonised grain remains and particular types of settlement, and hence assessing the degree to which the processing of agricultural products was centralised. Simply knowing where raw materials were obtained, where production took place, and where products were consumed does not tell us how these processes were organised and why. We cannot doubt that many activities are designed for physical survival, but survival is not achieved in stasis; we must reproduce ourselves physically, socially and, at the level of the individual, psychologically.

In this manner we develop different modes of reproduction. This is rarely, if ever a direct response to our environment; rather, it is a mediated response, but in interaction and production the medium will alter. In conversation the medium will be the voice, in work the raw materials and, ultimately, the product. Thus, the clay from which the pot is shaped is a medium for production, and the pot itself becomes a tool, mediating between the thirsty human agent and the water but also, in its particularity of composition, conveying traditional information, whether consciously or not.

In Europe the excavators of a classical sites may encounter varied architectural remains ranging from the grand public building, to the well appointed dwelling with a bath house and underfloor heating, to the craftsman's house and so on down to the barracks, the rural *vicus* and the isolated farmstead. Pooled, this information provides a rough guide to social hierarchy, and it is readily conceivable that a complete range of site types have been described. But until we know more about the frequency and distribution of those various types, and until we understand how the land in between structures was utilised, we cannot truly understand how they functioned together. We cannot provide a substantial argument to explain why the poor peasant has only the remains of the least fleshy portions of the animals he raised.

The question we must ask is "at whose behest or for whom are these activities being performed?" The whom may be the population of a territory, a select few or an individual.

4.4iiia   Social authority/ideological authority

Authority does not exist without the transmission of information, whether it be coercive or consensual. Only rarely will we re cover direct evidence of the transmission of authority in the archaeological record but we can see some of the products of subsistence and economic activity and certain features in landscapes which are strongly connotative of ritual activity.

In my classification of narrative levels I have separated into parallel strands the social and the ideological (*2.1*). This is not a false dichotomy, but the distinctions between the two classes are often blurred, and ultimately they are probably in all cases subsumed into an overall authoritative system. The division allows us to see more clearly the strands of narrative, even in circumstances where a single category of object may function in both classes. For instance, we are aware that in the classical world statuettes and other trinkets were distributed from centres sacred to particular gods, but whereas the act of purchase (if purchased they were) may have been prompted by piety, the act of selling may have had economic ends at least as compelling as the intention to reinforce a prevailing ideology. On the other hand we must account for the distinct macro-regional distribution patterns of objects with a very mundane form - axes - which have been fashioned from materials unsuited to practical use. Clearly the use of such an object is raised beyond subsistence, but should it be seen as currency or as the concretion of ideology?

In excavation we can discover configurations of material which appear to have no direct role in economic production but which, by the care devoted to their organisation, must have been invested with special significance either as objects in themselves or as the residues of a special activity. Much of this material would have been drawn from the mundane, but some of it would have been particular to a ritual function. Where material residues specific to ritual are comparatively rare it is necessary to look for recurrent constellations of artefacts which were used in this way. In looking at the way in which language enables communication V. N. Volnshinov[1] made a distinction between the lower limit of meaning as represented by a particular word "in the system of language or, in other words, ...a dictionary word", and its upper limit, which resides in the context of utterance (Voloshinov 1973, 102). These two limits of meaning are "*the technical apparatus/or the implementation of theme*" (Voloshinov 1973, 100), where *theme* is what is expressed in actual utterance.

Theme corresponds closely with what it is that an archaeologist is trying to discern in creating class F and, particularly, class G narratives. Although human systems change over time and space we are in effect seeking to recreate from incomplete material residues catalytic moments

---

[[1] V. N. Voloshinov was writing in the heady but dangerous 1920s, in the newly created Soviet Union. Armed with the then fresh perspective of Marxist theory his work ranged over philosophy, literature, language and even psycho analysis. In many respects he anticipated the European structuralism of the 1960s. He did not survive the decade. I am grateful to my former philosophy lecturer, Professor Frank Cioffi of the University of Essex, for introducing me to his work.]

of articulation. We are seeking the framework within which the human agent was thinking and externalising that thinking. Our project is not unlike that of Freud, who would often begin the processes of analysis, or dissolve blockages of utterance, by encouraging the patient to utter any seemingly unconnected words which came to mind. He would then look for associations between these artefacts of utterance, considering their frequency and intensity, and would attempt the reconstruction of particular scenes which had slipped from conscious memory. The moment Freud hoped to find was not one that laid out his subject's personal ideology, but expressed the origin of part of it. As archaeologists we rarely have the opportunity to describe an origin of ideology, rather we seek a moment in which an aspect of ideology is expressed, either of the instrumental, political kind or the self/social belief kind.

Although moments of class F narrative may be encapsulated and expressed in the perception of a particular configuration of artefact it is a level of interpretation which has an indirect relationship with the material. A past mode of subsistence, an economy, is only understood through the juxtaposition of functional, chronological and distributional interpretations. Through creating hypothetical units of synchronous activity we designate areas of ceramic production, of metals extraction, of pasture, of waste disposal, of dwelling, of ritual observance etc.

In general, interpretation of such units has depended on the notion of their interrelatedness, although even absolute proof of contemporaneity is not proof of association. But if we allow that some synchronous units were constituents of a system our goals will be first to describe and then explain the system. Various forms of systems analysis have been suggested, though rarely thoroughly applied. Gregory Johnson (1978, 89), identifying minimal organisational units, or sources, such as territorial, population, residential and activity units, highlights the transfer of information as a critical feature to be identified, presumably in the shape of material residues (the wires of David Clarke's Black Box). Arguing on the basis of maximised efficiency he identifies vertical and horizontal models of authoritative dissemination. He postulates a hierarchical chain with a pyramidal structure necessitated by limits to the functional capacity of each tier in the system. Horizontal organisation occurs where units are broadly self-authorising.

This mode of analysis has a limited concept of efficiency, and an unduly strong sense of *telos*. Implicitly shared elements of the perspectives of Henri Lefebvre (as expounded by Derek Gregory 1994, 382-387) and John Barrett show social reproduction to be a struggle to conserve which *progresses* only paradoxically in the effort to accommodate competing forces. If the body serves as a model for the organisation of "primitive" space this accommodating is a form of ingestion, a making-part-of which contrasts with systematisation. Giddens has characterised systematisation as the exercising of power in absence through channels of communication (Gregory 1994, 115-118). "System" is an abstract, historically and spatially (western) situated concept and explanatory models which depend upon it are limited in their application.

My contention is that survey should be regarded as an essential aspect of any attempt to reconstruct patterns of authority, since the scale and the character of distribution are as much elements of political and ideological expression as the "key" prestigious sites which have often been treated as the definite products of past human systems. But we do need to excavate. The safest association of objects in time and within space is achieved through digging. Catalytic moments of articulation are indeed manifested in particular monuments, but we need to know if those moments were co-occurring when we consider a complex such as the Dorset Cursus and the barrows situated around it. For this we need stratigraphy! We also need to dig settlement and industrial sites which can be shown to have existed during the period of a monument's construction and use the data to test the relationship between the structures of economy and belief. Only then will be begin to develop the analogies from which inferences about surface distributions may be made.

5: Babel

5.1 The problem of comparability

The aim of this chapter is to establish a means for maximising the value of regional surveys. To achieve this it is necessary to agree upon a structural language compatible with the varied and inconsistent terminologies and analytico-interpretative schemes which continue to be employed. There have been calls for "some form of compatibility in the presentation and interpretation of results" (Schofield 1991a, 117. See also Keay & Millett 1991, 133) but the circumstances in which individuals or groups commence a survey are frequently very ad hoc, and the methods employed will greatly influence and restrict the modes of presentation. However, whilst it is unrealistic to expect all new surveys to adopt a uniform lexicon, a categorising of project field and interpretive techniques will provide a medium through which it may be possible to assess the quality of planning, implementation and analysis of surveys. Such a scheme for assessment is vital if we are to create the meaningful dialogue which allows macro-regional narrative based upon accounts derived from a diversity of projects.

The particular structure of field and desk-based research, subsequent analysis and interpretation of a survey should be governed by its objectives, but all surveys will share certain general characteristics with other surveys, derived from their overall aims, as well as specific objectives which generate similarities. Project directors state routinely that the final shape of survey methodology and sampling strategy is determined by the specific nature of the terrain which is their object, and the resources available; all too often this has been an excuse for avoiding rather than addressing a particular problem. At present surveys can be divided into three general categories:

1) **Site Location Survey (SLS)**: The objective is to find and map areas of archaeological interest. Typically surface collection will be employed, sometimes with transcriptions of air photographs. Fieldwalking may be according to the whim of the walker, or to a plan, but it will not be theoretically-based. Chronologies of findspots or sites will be broad band, i.e. Neolithic, Bronze Age, Iron Age, Roman etc. The results of such a survey are not suited to providing a level of narrative beyond class C, but they may serve to stimulate further work in the region. They will make a useful addition to a sites and monuments record.

2) **Assessment Survey**: An assessment survey will employ some of the methods of site location survey, but in a planned manner. In broad terms there are two types of assessment survey. The first, which I shall term the **Heritage Resource Assessment Survey (HRAS)**, is directed towards known sites or monuments. Its aim is to monitor the condition of a site and to assess forms of risk to it. Its remit may well include consideration of means for conserving the site/region and controlling access to it.
The second, which I shall term the **Evaluative Assessment Survey (EAS)**, is directed towards areas at imminent risk from development where significant archaeology is suspected, but poorly understood, or where the area effected is so extensive that significant archaeology is likely to be encountered for reasons of sheer scale. Additional field methods are likely to include test pitting and trial trenching and, with increasing frequency, geophysical survey. At present EASs are geared towards discovering sites within a threatened region; rarely will their data support more than a class C narrative.

3) **Research Survey**: The research survey has as its objective the making of statements about past human activity derived from analysis and interpretation of data. It may be restricted to a desktop study, but most such surveys include fieldwork.
A feature which distinguishes the research survey from others is the extent to which the significance of the data is understood, or at least considered, during the analytical process. Thus, desktop researchers need to be aware of the limitations of the literature they use, while fieldworkers will take steps to determine whether rates of surface collection are due to discard, soil movement or other processes and so on.
The research survey should be aiming to achieve at least a class E narrative.

5.2 Categorization of regional research surveys

It is the research survey which is the object of this thesis. Under its banner are a host of different approaches both in terms of the selection and division of the landscape for sampling, the methods employed, and staging of the work. Modes of staging do not offer the criteria for effective differentiation between projects, although there are significant variations. It has been common for stratified random experiments in sampling strategy and surface collection intensity to be conducted during a preliminary stage. Equally, there are examples of "extensive" preliminary stages, followed by "intensive", focused survey of areas perceived as interesting. The opposite approach is to work intensively in areas known from the literature to be of interest within the survey's target periods, usually including some intrusive procedures, to create a comparative database which will enable swifter, better informed assessment of data recovered from the region in general, where less intensive methods are used.

It is rare for a project to be able to plan a staged approach and then simply to carry out each stage. Commonly, each stage, even parts of stages, require new applications for funding. For this reason, the stage sequence may be crucial; a preliminary or interim stage must not only demonstrate that there are significant target data, but that it is capable of recognising those data at a level of resolution which will facilitate more effective fieldwork and analysis during the next stage.

Although staging should be regarded as a benign aspect of project design there are often practical problems where a general reconnaissance has been conducted, using a longer sampling interval than that employed for later stages. A more pressing problem arises from in-the-field decision making about the significance of surface data. It remains common for fieldwalkers to use their "experience" to judge whether or not a scatter is worthy of collection or of detailed recording. The discovery, after one or two seasons, that the criteria employed by the project for the designation of "sites" are inadequate is a recurrent feature of reports, although I have yet to come across an instance where the directors have decided to completely re-survey an area. Consequently, the problem of data comparability occurs not only between surveys, but within them. One means for overcoming this problem is to adopt a limited but intensive preliminary phase aimed at creating a reference chronology, and at understanding examples of the relationship between surface and buried material. Analysis of the subsequent data should allow a more accurate assessment of appropriate methods and sampling frequencies needed to address the issues most pertinent to the aims of the survey.

5.2i Resource distribution classification

Susan Alcock introduced a very summary system of survey classification for evaluating the comparative usefulness of twenty-one projects which provided data for her super-narrative for Greece under Roman rule (Alcock 1993, 35-37; table 2). She prioritised the intensity of surface observation and the need for geomorphological research and rigorous methodology, distinguishing between line-walking surveys of upto 20m (group A) and less intensive, unsystematic (group B) and extensive procedures (group C). In the scheme outlined in this chapter group C surveys would be classified under Site Location Surveys, whereas A and B surveys would be given much more detailed classifications under their general Resource Distribution class, to which would be added Spatial resolution and Chronological resolution classifications. Variations in the intensity of work within stages are linked to the more general issue of the distribution of project resources in the field across the whole region, through the duration of the survey. Resource distribution is a useful criterion for comparing one survey with another, since it takes into account variations of intensity of work. A history of survey would show a general trend away from site location (SLS), towards the region as a continuum, to subregions set within the continuum. However, it would be rash to assume that one approach is actually superseding another. The adopted mode depends on the current state of technology and the availability of resources, and many examples of all three of these approaches could be found in use now. In subsequent discussion they will be treated as co-existing, rather than serial, modes of operation. But first these criteria must be revised to take into account the difference between their application as field and as analytical strategies. The tree in Figure 5.1 summarises the regional distribution classification.

The region is the whole area designated for survey. Each form of survey within the tree may be described in the following general terms:

1) **Regional, Multifocal (R/MF):** Rare, usually occurring as a survey of a region's literature. It may be oriented towards a particular topic, and there will be little or no attempt at even coverage or systematic sampling. An effective example is the combining of botanical evidence from existing and current excavations from an area of north east England (van der Veen, 1992). A very stimulating example is provided by work in and around the Dorset Cursus, combining an extensive literature with selective excavation.

2) **Regional, Multilocal (R/ML):** A project seeks to represent the region by survey work exclusively within localities which may vary in size (typically exceeding 50ha, and less than 10km^2), and which may be targeted to address specific archaeological problems within that region. Work within the locality may or may not be systematic. Although there is no theoretical objection to uniformity of coverage within the localities, in practice such surveys tend towards more intensive work around particular foci.

   a) **Regional, Multilocal, Unifocal (R/ML/UF):** Typically this variant of the R/ML is centred on a particular known site. It is usual for the site to occur within one of the localities.

   b) **Regional, Multilocal, Multifocal (R/ML/MF):** The optimum form of this survey would take the shape of intra-locality continual-type coverage, with one or more foci in each locality. That ideal is rare; the Danebury Environs Project, although nodding in the direction of a *regional* continuous model, eschewed methods of fieldwork which would have sampled a large portion of the region in favour of one large-scale geophysical survey (Windy Dido), and seven loci/foci of excavation, sometimes supported by preliminary geophysical survey of the site.

The great majority of research surveys are Regional Continuous (R/C); whichever smaller areal units receive special attention, there is treatment of the region as a whole by coverage of its total area, or by adopting a sampling strategy to represent probabilities or possibilities within it. Many of the surveys below are not fully Continuous in the explicit sense intended by Clarke and Foley, in that the majority of them sought to identify sites and did not attempt to use the full range of surface artefact density, which would include zero. However, most projects were at least aware that the site cannot be regarded as a wholly discrete entity, and that varying concentrations signify differing types or intensities of activity which contribute to an overall pattern in landscape space and time. Their deviation from continuousness may be regarded, and dealt with, as variance in methodological resolution. The continuous treatment of a landscape excludes any survey which is restricted to a review of the literature. In continuous survey the bulk of data collection will be at ground level, in the field.

3) **Regional, Continuous, Unifocal (R/C/UF):** A single focus (for instance, a substantial archaeological excavation, or series of excavations within a perceived occupation unit such as an estate or farmstead) is used to provide a

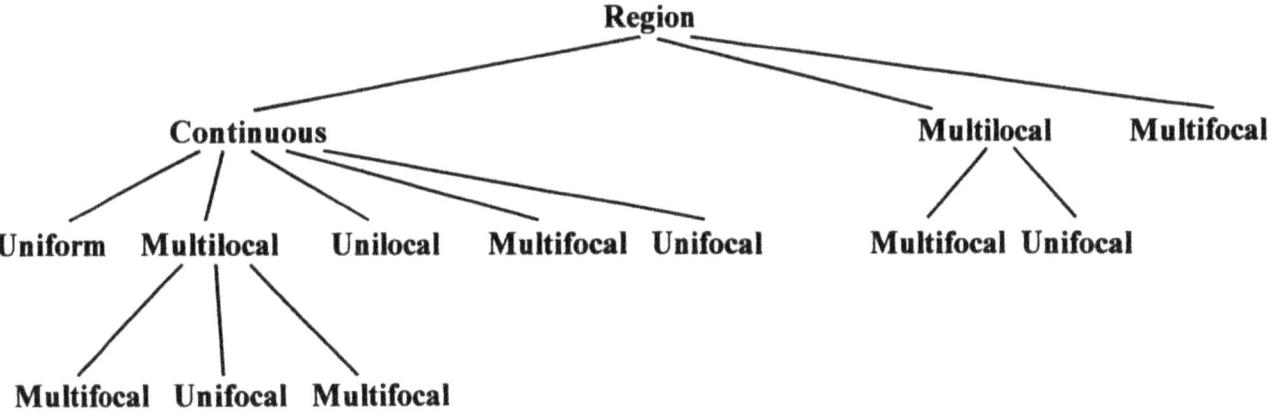

Figure 5.1. Resource distribution classification tree

chronological reference for the continuous survey. Alternatively, a continuous survey may be used to generate contextual information about the focus.

4) **Regional, Continuous, Multifocal (R/C/MF)**: Either a cluster of activity areas are used to provide a chronological reference for a continuous survey, or a continuous survey generates contextual information about the foci.

5) **Regional, Continuous, Unilocal (R/C/UL)**: Continuous survey either providing a context for, or gaining a comparative database from, a locality within which more intensive work is carried out. This form of survey may weight resources towards one or more foci within and/or outside the specified locality.

   a) **Regional, Continuous, Unilocal, Unifocal (R/C/UL/UF)**: A form of R/C/UL where resources are weighted towards a particular focus, either for the purpose of obtaining a regional type series, or because the project is intended to provide a context for the focus.

   b) **Regional, Continuous, Unilocal, Multifocal (R/C/UL/MF)**: A form of R/C/UL where resources are weighted towards two or more foci, either for the purpose of obtaining a regional type series, or because the project is intended to provide a context for foci, which are considered to have had principal roles in characterising the region during the target period. The Tarragona survey is a good example, where a single urban locality is informed by, and provides a comparative database for, a R/C survey within which certain concentrations of data are the foci of special attention, including geophysical survey and excavation (Carreté et al. 1995).

6) **Regional, Continuous, Multilocal (R/C/ML)**: Continuous survey either providing a context for, or gaining a comparative database from, two or more localities within which more intensive work is carried out. This form of survey may weight resources towards one or more foci within and/or outside the specified localities.

   a) **Regional, Continuous, Multilocal, Unifocal (R/C/ML/UF)**: A survey in which the continuous approach at regional and local level provides layered contextual information about a particular known site, or where a single site is used to provide a chronological type series for a region within which specified localities are treated more intensively to increase the resolution of distributional information.

   b) **Regional, Continuous, Multilocal, Multifocal (R/C/ML/MF)**: A survey in which the continuous approach at regional and local level provides layered contextual information about particular known sites, or where certain known sites are used to provide a chronological type series for a region within which specified localities are treated more intensively to increase the resolution of distributional information.

7) **Regional, Continuous, Uniform (R/C/U)**: The distribution of project resources is applied uniformly, either across the whole region or within a sampling strategy, random or stratified.

Some current opinion regards the R/C/U as an ideal, providing it is possible to apply a sufficiently frequent sampling interval; however, it is difficult to see how this scheme can meet the need for chronological or functional resolution. Other methods are not as susceptible to generalising statistical analysis, but a well thought out R/C/ML variant can be adapted to it, enhancing chronological resolution through excavation, and improving functional resolution by looking for patterns visible from more intensive work at the locality level.

These categories of regional research survey deliberately eschew terms which are associated with measured space. Instead they reflect how the archaeological surveyor conceptualises space. The scheme provides a framework within which individual surveys can be grouped in a meaningful manner, but as such it is no more than the structure into which technique, measured space and time must be introduced to develop an effective mechanism for comparison. In Chapter 3 a language of space has been outlined (*3.2*), and a range of techniques listed and discussed (*3.3*). The work of that chapter will ease our progress towards a means for assessing the relative degrees of spatial resolution and reliability between surveys. The issue of time cannot be fully separated from space (*4.3*) but chronological information can act as an interrogator of space. Simply, when we hypothesise about social or cultural systems and entities within a particular landscape we must have information concerning the probability of coexistence and succession, at least, to show order in that space.

Only by breaking down the region or landscape into groups of component characteristics will we begin to understand the significance of the distribution of data, whether artefactual, chemical or structural. In effect these characteristics are the equivalent of strata, whether or not they have been conceived as part of the research plan at the outset. The difficulty with such strata is that all are susceptible to change over time, some within comparatively short periods. Hydrology and soils can alter radically within a few centuries, and even topography will vary significantly over the course of a few millennia. The viability of an area may be affected by volcanic activity and earthquakes and, as Mediterranean surveys have shown, variations in sea level, even tectonic movement, have had considerable impact on the amount of available land, and the character of the landscape through the creation of deeply incised, steep valleys or broad alluvial plains within archaeological rather than geological time (Rapp & Kraft 1994, 69-70).

This is not to suggest that strata are too unreliable to be applied; the longevity of some of those selected is likely to exceed the duration of the period of interest to many projects. Where this is not so the archaeologist is bound to recreate a succession of landscape snapshots reflecting the development, expansion, contraction or disappearance of strata onto which artefact distributions are mapped. Good, very detailed, examples can be found in any of the Fenland surveys of Britain's East Anglia, where cycles of inundation and peat development have lead to extreme variations in the availability and quality of land.

From the analytical point of view a stratum has the status of independent variable. A set of regional strata provides the potential for observation of regularities in artefact distribution, whether structural or portable. The regularities in the distribution of dependent variables are the foundations for the inferential process, when we offer explanations for why a particular distribution occurs. However these explanations will be in general terms.

Refinement of explanation occurs as the potential for specificity of variables increases. Direct associations of portable artefacts with structural features, other portable artefacts, or details within the landscape (such as "aspect" or "proximity to water") will determine the potential for diagnosis of time, function and group affiliation. These might all be described as variables of explanation, whether independent or dependent. As such, they are distinct from another set of independent variables - those which show the effect of the act of collection or observation upon the perceived distribution of data.

5.2ii   Spatial resolution classification

For many earlier research surveys, in which the principal technique was surface collection, spatial resolution was a straightforward issue. Project directors accepted that the goal was simply to cover as large an area as possible with the greatest interval between walkers which would allow the location of significant sites. During the 1980s, especially, a debate developed from the recognition that a site of any size can provide valuable social and economic information. From which do we learn more; the grand house which ran the estate, or the shepherds cot from which flocks were husbanded, providing the estate with tradable commodities? Unless it can be shown that by focusing on high status elements in the landscape we can necessarily infer a comparable amount of information about lower status elements, it is impossible to avoid arguments over political perspective if we chose to treat lower, usually less perceptible elements, as insignificant.

It has been shown conclusively that the rate of locating smaller findspots is heavily influenced by the interval between lines (*3.3ixa*, above). Surveyors must either: accept multiplied-up estimates of low frequency findspot distribution, specify particular classes of sites as targets or resort to much more intensive techniques. If the latter approach is preferred survey directors will either have to cover smaller areas, increase their labour forces or lengthen the duration of projects. Hodder and Malone (1984, 127) suggest that a planned mixture of extensive and intensive techniques should be employed, although in Stilo the results from the intensive procedures were far more informative than those from the extensive. It is not the aim of this section to arrive at what should be done, rather to enable the classification of what has been done. From this designers of future projects should be able to make rough assessments of the quality of information their fieldwork will generate given the resources available to them.

As has been amply illustrated in Chapter 3, fieldwalking is far from being the only technique of survey, and spatial resolution is just as influential over the results from other techniques. However, in establishing a classificatory system we need only deal with field techniques which combine chronological with distributional information. Thus, geophysical prospection may offer plenty of distributional information, but without supplementary techniques chronology is usually weakly resolved (although Palmer 1984 has shown a chronologically significant morphology for aerial photography, the data of which can be compared with those from geophysics). Only techniques based on the recording or collecting of artefacts offer this necessary combination. Excavation and trial trenching are supplementary techniques applied focally, so we are left with test pitting, shovel testing and fieldwalking to determine spatial resolution. Any of these may be applied locally or regionally, either exclusively or in combination, according to the demands of a particular terrain.

In many regions simply applying fieldwalking cannot be the basis for describing spatial resolution, whether aiming for total coverage or a percentage of sampled areas, because of variation in the visibility of artefacts caused by erosion and other factors. In such cases either geomorphological research will give an indication of what statistical significance should be attributed to surface variation, or ploughzone or intrusive techniques should be employed in areas where conditions make surface collection inappropriate. It is preferable to establish as clear a correlation as possible between the results of fieldwalking and auxilliary methods.

For clarity I shall consider first the situation where ground conditions after widespread cultivation provide good visibility, and we may assume that the plough has a uniform effect upon archaeological contexts, so yielding a consistently representative crop of surface artefacts. If the surveyor's plan is to view every bit of the surface, with an equal expectation of seeing material over each bit, walkers will need to be closely spaced. In Boeotia it was assumed that there was "an effective visual range of 2.5 m for each walker on either side" (Bintliff & Snodgrass 1985, 130). That may be true, but it is certainly not a uniformly effective range. A person walking is most at ease when he or she can look straight ahead. When the field of focus is only 2 to 3m from their eyes a lateral span of only a little over 1m is likely to receive a consistently high level of attention. To achieve full coverage on that basis fieldwalkers would have to stand adjacent to each other, and only 1m apart. But is it necessary to view the whole surface?

Only in an environment such as that in the Amboseli, where erosion has removed the soils which have formed since deposition, is it possible to view a very high proportion of the discarded durable artefacts. Otherwise, when a fieldwalker views the surface of a cultivated field he/she is merely seeing a percentage of the material in the ploughzone. Bearing in mind the homogeneity of the ploughzone, and the well-attested dispersal of artefacts from their point of horizontal origin only a few episodes of cultivation will have created a halo over a considerably larger area than that of the deposited concentration of artefacts. Of course, a once-used hearth of 1m diameter may be the source of very little. Even if a whole cooking pot had been abandoned, generating a few large sherds subsequently broken into around 20 smaller ones of which perhaps 5 are on the surface at the time of fieldwalking, there is a good probability that only one sherd will be available for collection! On the other hand if a hearth, around which three or four men had sat knapping flint, was me centre of some 100 debitage chippings in a rough circle of 3m diameter the experimental literature suggests that there would be around 5 surface flakes within a circle of 15 to 20m diameter. In such a case variations of plotted densities for lines only 1m apart might locate discrete areas of activity. One or two lines set at 10m intervals will bisect the circle of dispersal, but may not locate enough flakes, if any, for an ADABS to be noted.

Is this a problem? If such a flint scatter is representative of a hunting party's temporary camp how important is it that we should know about it? If hunting was an essential component of a population's subsistence we need to be able to see the evidence for it, but we do not necessarily need to find the particular hearth itself. We can consider evidence for the activity in general, just as Foley did in the Amboseli, in a manner which enabled to him to construct a convincing argument for a particular mode of subsistence derived from ethnography and a clear view of land holding capacity. Of course, Foley was not looking for *sites*, rather he was associating an activity with an environment where formal division of the landscape, its tenure, and overarching social organisation were not issues. On the other hand his data would not have contributed to a debate concerning division of labour or exchange because, although each collection unit was searched intensively, the units were extremely widely spaced!

Line walking in 1m wide strips set at 20m intervals would almost certainly be capable of producing similar generalised patterns in flint distribution. Furthermore, where ceramics were present, it would very probably show evidence of even quite low-level exchange, and would locate a substantial number of activity foci in the form of artefact concentrations. This scale of investigation should show some variation in function and status, hut even after repeated ploughing it is likely to miss source concentrations 4m across, comprising 100 artefacts. Whereas this would not be a problem for acquiring the generalised information sufficient to denote a mobile mode of subsistence, it may cause difficulties where interpretation of a targeted period seeks to demonstrate relative status and function in an elaborate economic system, within which occupation of land was more sedentary. On the other hand, sedentary lifestyles are likely to generate larger or denser concentrations of rubbish and so be more distinctly visible.

As noted above, recent literature suggests a maximum effective visual span of around 2m (*4.3i* above). Personally, I would prefer to reduce it to 1m, and that outlined by ropes in the field. However, since the majority of surveys allow the larger figure it is probably appropriate to treat that as a basis for a classificatory system. Thus if the whole of a segment or sample area of a stratum is covered at 5m intervals, I shall assume that 40% of the surface has been searched, and if the intervals are at 50m, that 4% has been searched.

Where fieldwalking is an inappropriate method and shovel testing is used, the frequency of sampling and either the soil volume or the surface area of each sample must be taken into account. If the test comprises a surface area of $1m^2$, taken at every 10m along a line, with lines at 10m intervals, 1% of the area has been sampled. However, this figure is not comparable with a 1% fieldwalking area, since it comprises soil to the depth of the plough which has been sieved. If we assume that around 5% of the artefacts within the ploughzone are on the surface at any one time we can infer that as a general rule a surface sample is approximately 20 times less representative than a sieved sample. Applying the principle of ploughzone homogeneity, we can go on to say that a $1m^2$ provides information which is approximately equivalent to a 20% surface sample[1]. Of course, the smaller the amount of data collected (the sample *size*), the less reliable this correlation becomes. If the aim of a survey is to recognise sites this equation would become much less reliable as the interval between samples increased, but it should be effective where generalised information is required. If shovel testing is conducted by collecting a specific volume of soil no such equation can be made since the proportion of soil volume sampled will vary against the percentage of area sampled because of variations in ploughsoil depth.

Before enumerating classes of spatial resolution we should

distinguish between three different types of spatial resolution. The first type concerns the frequency and extent of surface observation; the second type concerns the frequency of collection; the third type concerns the frequency of representation (sample points), either on maps, or as grid referenced groups of finds. Many surveys carry out surface observation in a manner which does not reflect the collection unit (see *3.2* above), observing the surface in multiple lines within a single field, then amalgamating all the finds from the field, which become a single sample point. Typically, in Mediterranean surveys, these fields are estimated to have an approximate area of between 1 and 2ha. It would be a very straight forward matter, if more labour consuming, to increase the frequency of sample points. This leads on to the matter of mapping results and the subsequent analysis. The argument used against increasing the frequency of sample points is that the absolute sample size must be large enough to be statistically manipulable, and the smaller the area observed the less likely the artefact sample is to be sufficient. This is something of a non-argument, because if the sample is too small a larger number of collection units can be amalgamated under a single sample point. Indeed, there is no reason why more than one sample point scale should not be used on a survey's distribution maps since it is commonly the case that the frequency of artefact occurrence will vary according to the time of their deposition. Whilst inter-period comparisons must be presented at the same scale, intra-period studies of activity distribution should be conducted at the finest resolution possible. This will increase the amount of labour time required, but adept use of a good spreadsheet package should avoid an excessive burden.

Where an archive exists with clearly described references for all collection units, a survey's spatial resolution may be described on that basis. If only the post analytical sample points survive resolution will be limited by intervals between them. Before proceeding to outline categories it is necessary to consider what may be achieved by the various possible scales of sample point. It has already been suggested that for certain forms of low density scatter derived from mobile subsistence systems the generalisation achieved by small scale mapping may be an advantage. However, for target periods during which more sedentary modes of subsistence were pursued it may be possible to map signs of burnt material resulting from long term repetition of activities, or structural debris, so increasing the likelihood of being able to estimate settlement size, and drawing sharper lines between different activities ranging from dwelling, through agriculture to specialised production.

In the following scheme spatial resolution is described by two numbers (Table 5.1). An Arabic numeral refers to the intensity of fieldwork, whilst a lower case roman numeral refers to the frequency of plotting or grid reference listing in the archival or published records.

5.2iii   Chronological resolution classification

Undoubtedly, in the present state of typological knowledge, particularly lithic and ceramic, this is the most difficult and most controversial mode of classification, and many practitioners may regard it as an inappropriate one, pointing out the extreme variation in resolution from one period to another. It is because the treatment of time varies not only from period to period, but from survey to survey that it is necessarily incorporated into the scheme. In effect the investigation of time is limited by a resource distribution choice between either work in the field and artefact seriation, or the latter and another aspect of analysis. All too frequently it is clear that there has been no attempt to improve upon existing ceramic and lithic series for a region.

In a majority of the examples from the Mediterranean there has been a lazy acceptance that coarse wares have little diagnostic value, to the extent that a site may be recognised but not dated. More recently some surveys have endeavoured to create their own series, or to enhance existing series. Clearly, in Britain for instance, we remain a long way from being able to produce a detailed diagnostic sequence for the Bronze Age which could emulate that of the Roman period. In part this is due to a lack of excavated deep, full, stratigraphy for a period where settlement was comparatively ephemeral.

One of the criteria for distinguishing a research survey from a site location survey is precisely the expectation that the former will produce much more than a gazetteer which may inform future work: it should strive to provide an archaeological history of the selected region. For the reasons given above (*4.2*) such a history can be improved upon only by increased chronological specificity.

In measuring space we apply arbitrary scales, with arbitrary beginning and end points which establish a set of values independent from the matter of research, but with points of reference in the landscape which are not only definable, but which remain observable. Modern absolute dating techniques provide us with landmarks in time which, although less reliable than landmarks in space (excepting where high quality dendro-dating is feasible), at least give us rough operant parameters for the application of an arbitrary scale. In many cases scales will have to vary from period to period because of the variation in chronological specificity, and the system of classification must reflect that range.

Absolute dating of the kind provided by some dendrochronology is likely to remain rare; as a rule the issue is one of assessing probable contemporaneity. We must decide what represents a reasonable optimum span of time. We must accept that close sequencing of transient camps by identification of technological variation will not be possible for the foreseeable future, but what is the shortest span of time during which there is perceptible and definable variation in Romano-British or Black Mesa ceramics? Without allowing for residuality I would suggest that span might be around a generation in length or, with a little leeway, twenty-five years.

In Britain and Europe it is common enough for Late Iron Age, Roman, Medieval and Post Medieval assemblages to be datable within 50 years, and a larger group to within a century. For other periods, and elsewhere, spans of 200 and 500 years may represent the smallest definable units of time, but for the European Mesolithic and even the Neolithic,

([1] I am aware, of course, that in most cases ploughzone surveys do not sample all the ploughsoil within a 1m2 of surface area. Thus for a ploughzone of depth averaging 30cm, one would have to sample 300l of soil. If only 30l of soil are sampled it may be argued that the shovel test has an approximate equivalence to a 2% surface sample.)

longer spans may be necessary. Table 5.2 summarises the scheme.

In a survey report the reliability of chronological resolution will be dependent upon the character of the evidence, but also upon the use to which the evidence has been put. The aim of this scheme is not to show which elements are common to a particular phase, rather it is to address the related, but more specific issue of what is the probability that different elements within the region actually coexisted.

If a survey is able to produce substantiated class 1 data, and the object of research is a comparatively sedentary society, we can reasonably expect a high incidence of contemporaneity of activity and, hence, potential for interaction between elements represented by finds.

Such evidence might be in the form of exceptionally closely dated ceramics, such as those from the Black Mesa or, less reliably but more frequently, in the shape of a ceramic series, the constituents of which can be shown to have been present at various stages within a temporal span. From this there may be a probabilistic assessment of actual contemporaneity, preferably with an estimate of reliability. This latter approach is likely to be of greatest value in raising chronological resolution to a higher level. How might it be effected?

First of all there is a need to come to a closer understanding of the existing data. Hodder and Malone (*4.2*) have demonstrated that variations in fabric compositions can be sensitive chronological indicators, suggesting that more use should be made of thin sections (a group in Lincoln is already working to create a centralised data base of thin sections). Where fabric or form indicators exist effort should be made to link every definable attribute to ^{14}C, TL or dendro dates, and these should be made available from a central source (the Internet comes to mind). Taking the example of British prehistoric pottery, a large body of information already exists - it simply has not been centralised and made generally available. Of course there is much local variation, but the degree of diversity of the vessels would be highlighted by such a project.

Once such a database has been compiled there is an opportunity to begin a planned programme to generate hypotheses about the probability of synchronicity of the various components of distribution patterns. Quite simply, we will have a chronological span of possible currency for each particular type of pot, within which there is like to be a modal pattern. The same approach can be adopted for lithics, although the probability of co-occurrence is almost certain to be much lower.

5.3 Categorisation and conclusions

The stated aim of this book was to explore the link between sampling strategies and their subsequent narratives. In Chapter 2 a simple mode of narrative classification was proposed. In this chapter the implications of field and analytical techniques and concepts explored in chapters 3 and 4 have been used to classify surveys according to their resource distribution, their spatial resolution and their chronological resolution. Table 5.3 classifies some of the surveys which have formed the body of this research.

The table does not show a comprehensive list of the projects referred to in this thesis, but it probably is an apt representation of the range of research surveys conducted by North American and European archaeologists in three continents. This latter bias reflects my limited linguistic skills and also something of the pattern of recent colonial history. Large areas of Asia have preferred to develop there own committed archaeological programmes, often in ideological opposition to perspectives which prevail in the western world. Consequently, my treatment of the topic has been framed within the particular range of western debate.

The number of surveys in the table has also been limited by the amount of published information available for individual surveys. It often takes several years for the final report of a survey to appear, so I have had to glean information from preliminary or topic-specific accounts constrained by the space available in journals or thematic anthologies. In these contexts papers have often included either a narrative overview, or a view of an aspect of methodology, neither of which has been sufficient to assess all of the resource distribution, spatial and chronological classes of the survey. I have had to exclude some very important surveys from the table because at the time their reports were written a particular class of information was not recognised as significant. A common example is the description of

1		A minimum of 10% of an area's surface is surveyed at a minimum frequency of 1m in 10m either latitudinally or longitudinally
2		A minimum of 4% of an area's surface is surveyed at a minimum frequency of 1m in 25m either latitudinally or longitudinally
3		A minimum of 2% of an area's surface is surveyed at a minimum frequency of 1m in 50m either latitudinally or longitudinally
4		Less than 2% of transect/segment's total surface area is surveyed
i		Mapped in blocks of upto 10m^2
ii		Mapped in blocks of upto 25m^2
iii		Mapped in blocks of upto 50m^2
iv		Mapped in blocks of upto 100m^2
v		Mapped in blocks exceeding 100m^2
vi		Spot-mapped only

Table 5.1. Classes of spatial resolution

fieldwalking as "total coverage". Bettinger went into considerable detail about his sampling methods in Owens Valley, and states that he covered the whole surface of each of his 500 x 500m segments. Does this mean that he respected the most stringent limits of visual span by walking every available 1m line within the segment? Or did he think that a line at every 5 or 10m would suffice? Without that specification we have insufficient information about spatial resolution.

Under the scheme for resource distribution classification proposed in 5.2 there are eleven possible regional research survey types under three general headings: *Continuous*, *Multilocal* and *Multifocal*. The only surveys not described as continuous are those around Cranbome Chase and in the Libyan Valleys. Although in Libya the areas in and around *wadis* were treated in that manner, the decision to work at specific locations was determined by targeting certain classes of highly visible sites. The report of the experiments in

1	Units upto 25 years
2	Units from 26 to 50 years
3	Units from 51 to 100 years
4	Units from 101 to 200 years
5	Units from 201 to 500 years
6	More than 500 years

Table 5.2. Classes of chronological resolution

Survey	First year	Last year	Resource distribution	Spatial resolution	Chronological resolution
Amboseli National Park	1974	1978	R/C/U	4vi	6
Biferno Valley	1974	1978	R/C/MF	2vi	4-6
Acconia	1974	1979	R/C/MF	1vi	6
East Hampshire	1977	1978	R/C/U	3vi	5-6
Cranborne Chase	1977	1984	R/MF	1i-(?)vi	4-6
South Dorset Ridgeway	1977	1984	R/C/ML	1iii	6
Boeotia	1978	1982	R/C/U	1(?)-2vi	4-6
Stilo	1979	1980	R/C/UL	4vi	5-6
Southern Argolid	1979	1982	R/C/U	1v-vi	3-6
Libyan Valleys	1979	1989	R/ML/MF	2-?	4
Metaponto	1981	1982	R/C/U	1	1-?
Maddle Farm	1981	1983	R/C/ML	2iv	4-6
Fenland Project	1981		R/C/U	3vi	5-6
Neo-Thermal Dalmatia	1982	1986	R/C/ML	2v-3v	4-6
North Keos	1983	1984	R/C/ML	2ii-v	4-6
Nemea Valley	1984	1989	R/C/UL/MF	2i-iv	4-5
Maiden Castle	1985	1986	R/C/UL	1iii-vi	6
Tarragona	1985	1990	R/C/MF	1v	4-5
Phlius	1986	1986	R/C/U	2iv	4-5
Patras	1986		R/C/ML	2i-3vi	4-6
Africa Proconsularis	1987	1989	R/C/ML/UF	2iii	2
Kavousi	1988	1990	R/C/MF	2iii	4-6
Shapwick	1988	1995	R/C/MF	2vi-1ii	5-6
Thy	1990		R/C/MF	2iv-1i	5-6

Table 5.3. Summary categorisation of a selection of surveys

surface collection and analysis from Stilo focus on just one locality, so in the absence of information to the contrary it has been treated as an R/C/UL, as has Maiden Castle.

We are left with three main variants: the R/C/ML, the R/C/U and the R/C/MF. Five surveys have adopted the first category, and a sixth a variant of it (R/C/ML/UF). Their spatial resolution for collection varies from 1 to 3, and from i to vi for mapping. Eight surveys have attempted R/C/U coverage. That does not necessarily mean that they have attempted to cover the whole surface (although at Phlius, for example, that has been attempted), rather it has been sampled randomly or systematically, not according to judgement. Collection resolution varies from 1 to 4, and mapping from iv to vi. The continuous multifocal approach was adopted by six of these surveys, with collection resolution varying from 1 to 2, and mapping from i to vi.

Chronological resolution for the R/C/ML surveys similarly varies from 4-6, with the exception of the Africa Proconsularis project which achieved class 2 resolution. Disappointingly limited use was made of this information since the largest scale mapping was presented in broader chronological bands. The R/C/MF has obvious attractions. The selection of a few small areas within a generally continuous sampling pattern might be expected to yield good chronological sequences, and better resolved, higher probability, synchronic distribution plots. The table shows that chronological resolution varies from 4-6. Some of the periods covered are earlier prehistoric, but some are also classical, for which better resolution might be expected. The chronological resolution for R/C/U surveys also varies from 3-6, with one extraordinary exception, that of Metaponto, for which it was claimed that class 1 might be possible (Carter & D'Annibale 1985, 150).

From this summary it would appear that the surest means for discrimination between surveys is on the basis of spatial resolution, making the R/C/MF the best performer. Those which have published final reports are the surveys of the Biferno Valley, Acconia, and Tarragona. All three have strong points in their favour, but the Tarragona account is thin on detail when it comes to describing the lives of people in a classical period. In contrast the Biferno project lacks sufficient detailed evidence to support a compelling narrative and the degree of spatial resolution varied. Ammerman was able to generate valuable insights into the means of subsistence and exchange in Acconia for a comparatively remote period (Ammerman 1985, 93-102), although the legitimacy of his wider conclusions might be doubtful.

Five of the six R/C/ML variants have appeared as final publications. Of these Peter Woodward's histories of the Neolithic and Bronze Age of the South Dorset Ridgeway display a tendency to allow narrative to outstrip the environmental data, but in other respects there is a strong correspondence with the evidence. In contrast, the Africa Proconsularis project failed to capitalise on a wealth of information, and it would be quite possible to attempt a more ambitious account on the basis of the data Dietz has published! Vince Gaffney and Martin Tingle made excellent use of local and regional data around Maddle Farm, within a macro-regional context to present a very accomplished socio-economic picture, but it would have benefited from greater chronological resolution and more environmental data. Much the same can be said of the account of Keos, although greater chronological resolution has added power to certain parts of the narrative.

In many cases the level of narrative is limited less by the quality and amount of the data than by the perspectives which research teams bring to a project. An exception is provided by the relationship between the methods and the story on Cranborne Chase, where the digging of monuments was a necessary part of establishing very local synchronies for getting into the mindset of Late Neolithic and Early Bronze Age agents. More commonly practitioners of the most rigorous and well thought out empirical surveys seem to be bound too tightly by the burden of proof, stunting their ability to speculate creatively. Very often this seems to be the consequence of a compartmentalised, static approach to a project, in which the results of a particular archaeological technique, notably fieldwalking, are given too much precedence. This pattern occurs most prominently in surveys which have given insufficient attention to the changes in the environment, natural and anthropogenic.

A well-founded story requires some intensive sampling and control techniques applied within subsets of the region, but at least as important seems to be the use of *variety* of techniques. John Bintliff has called for a "battery approach of overlapping techniques" (Bintliff 1997a, 238) in his pursuit of Boeotia's rural sites and detailed population distributions, but increasingly his database seems to be generating numerical abstractions rather than meaningful stories. If we can identify soil movement, land division, crop regimes, animal husbandry etc., we add more dimensions to the human figure, and the reader is better placed to envisage the agent in a working relationship with her or his landscape. We cannot understand the division of social classes and communities without the landscape and climate, since for most of archaeological time humans have striven to extract the means for life within it, have formed there understanding of life within it, have developed loyalty to it, and have fought to possess it. The most compelling regional narratives are those which have tried to convey how the peopled landscape works. The Southern Argolid survey published distribution maps which failed to point out the areas which had not been surveyed; in the Biferno Valley sampling was unevenly distributed and of variable intensity; and in the Libyan Valleys judgement sampling was applied; yet the reports of these surveys are not only the most approachable, but also the most convincing human histories because the subject is integrated into the landscape.

The categorising of surveys does provide some means for discriminating between "good" and "bad" surveys, but it also shows that a variety of approaches can produce results of comparable quality. It provides the survey archaeologist with a conceptual scheme he or she can use to break down a dauntingly large landscape into more manageable proportions by facilitating considerations of where both field and analytical resources may best be focused to address the questions that arise from a variety of different (ideological) perspectives.

Bibliography

Alcock, L. 1972. *'By South Cadbury, is that Camelot...': Excavations at Cadbury Castle 1966-70*. London: Thames & Hudson.

Alcock, L. 1980. The Cadbury Castle Sequence in the First Millennium B.C. *Bulletin of the Board of Celtic Studies* 28.4, 656-718. Cardiff: University of Wales.

Alcock, L. 1995. *Cadbury Castle, Somerset: The Early Medieval Archaeology*

Alcock, S.E. 1991. Urban Survey and the Polis of Phlius. *Hesperia* 60, 421-463.

Alcock, S.E. 1993. *GraeciaCapta*. Cambridge: Cambridge University Press.

Alcock, S.E; Cherry, J.F. 1996. Survey at Any Price? *Antiquity* 70, 207-211.

Allen, M.J. 1991. Analysing the Landscape: a geographical approach to archaeological problems In *Schofield, A.J. 1991*, 39-57.

Ammerman, A.J. 1985. *The Acconia Survey: Neolithic Settlement and the Obsidian Trade*. London: Institute of Archaeology.

Ammerman, A.J. and Feldman, M.W. 1978. Replicated Collection of Site Surfaces. *American Antiquity* 43, 734-740.

Arnoldus-Huyzendveld, A., Gioia, P., Mineo, M. and Pascucci, P. 1995 Preliminary Results of the Malafede Survey, 1990-1992. In *Christie, N. 1995*, 47-56.

Aston, M.A. and Costen, M.D. (eds). 1992. *The Shapwick Project: A Topographical and Historical Study. The Third Report*. University of Bristol Bristol.

Aston, M.A. and Costen, M.D. (eds). 1994. *The Shapwick Project: A Topographical and Historical Study. The Fifth Report*. University of Bristol Bristol.

Aston, M.A. and Gerrard, C.M. (eds). 1995. *The Shapwick Project: A Topographical and Historical Study. The Sixth Report*. University of Bristol Bristol.

Barker, G. 1995. *A Mediterranean Valley: Landscape Archaeology and Annales History in the Biferno Valley*. Leicester: Leicester University Press.

Barker,G. 1995a. *The Bifemo Valley Survey: The Archaeological and Geomorphological Record*. Leicester: Leicester University Press.

Barker, G. 1995b. Landscape Archaeology in Italy - Goals for the 1990s. In *Christie, N. 1995*, 1-11.

Barker, G. (ed). 1997. *Farming the Desert: The UNESCO Libyan Valleys Survey. Vol 1: Synthesis*. Tripoli & London: UNESCO.

Barker, G; Hodges, R. (eds). 1981. *Archaeology and Italian Society: Prehistoric, Roman and Medieval Studies*. Oxford: BAR (IS 102).

Barker, G; Jones, B. 1985. Investigating Ancient Agriculture on the Saharan Fringe: UNESCO Libyan Valleys Survey. In Macready, S. & Thompson, F. 225-241.

Barker, G. and Lloyd, J. 1991. *Roman Landscapes: Archaeological Survey in the Mediterranean*. London: British School at Rome.

Barnes, GL; Okita, M. 1993. *The Miwa Project: Survey, coring and excavation at the Miwa site, Nara, Japan*. Oxford: BAR (IS 582).

Barrett, J.C. 1994. *Fragments from Antiquity: An Archaeology of Social Life in Britain, 2900-1200BC*. Oxford: Blackwell.

Barrett, J.C., Bradley, R. and Green, M. 1991. *Landscape, monuments and society: The Prehistory of Cranbome Chase*. Cambridge: Cambridge University Press..

Barrett, J.C., Bradley, R. and Hall, M. 1991. *Papers on the Prehistoric Archaeology of Cranbome Chase*. Oxford: Oxbow (11).

Barrett, J.C., Freeman, P.W.M. and Woodward, A. 2000. *Cadbury Castle, Somerset: The later prehistoric and early historic archaeology*. London: English Heritage.

Bartram, L.E., Kroll, E.M. and Bunn, H.T. 1991. Variability in Camp Structure and Bone Food Refuse Patterning at Kua San Hunter-Gatherer Camps. In *Kroll, E.M. and Price D.T. 1991*, 77-148.

Batovic, S. and Chapman, J.C. 1985. The 'Neothermal Dalmatia' Project. In *Macready, S. and Thompson, F.H. 1985*, 158-195.

Baxter, M.J. 1994. *Exploratory Multivariate Analysis in Archaeology*. Edinburgh: Edinburgh University Press.

Bell, M. 1983. Valley sediments as evidence of prehistoric landuse on the South Downs. *Proceedings of the Prehistoric Society* 49, 131-142.

Bell, M. 1990. *Brean Down: Excavations 1983-1987*. London: English Heritage.

Bell, M. and Boardman, J (eds). 1992. *Past and Present Soil Erosion: Archaeological and Geographical Perspectives*. Oxford: Oxbow (22).

Ben Lazreg, N. and Mattingly, D.J. 1992. *Leptiminus (Lamta): a Roman port city in Tunisia, Report no. 1*. Michigan: Journal of Roman Archaeology supplementary series 4.

Bettinger, R.L. 1977. Aboriginal human ecology in Owen's Valley: prehistoric change in the Great Basin. *American Antiquity* 42, 3-17.

Bettinger, R.L. 1979. *Multivariate Statistical Analysis of a Regional Subsistence-Settlement Model for Owens Valley*. American Antiquity 44, 455-470.

Binford, L.R. 1978. Introduction to Nunamiut Ethnoarchaeology. In *Binford, L.R. 1983*, 179-194.

Binford, L.R. 1981. Middle-range Research and the Role of Actualistic Studies. In *Binford, L.R. 1983*, 411-422.

Binford, L.R. 1982. Meaning, Inference and the Material Record. In *Binford, L.R. 1983*, 57-62.

Binford, L.R. 1982. Objectivity-Explanation-Archaeology-1981. In *Binford, L.R. 1983*, 45-55.

Binford, L.R. 1982. Some Thoughts on the Middle to Upper Paleolithic Transition. In *Binford, L.R. 1983*, 423-433.

Binford, L.R. 1982. Working at Archaeology: The Generation Gap - Reactionary Arguments and Theory Building. *Binford, L.R. 1983*, 213-227.

Binford, L.R. 1983. General Introduction to For Theory Building in Archaeology. In *Binford, L.R. 1983*, 31-43.

Binford, L.R. 1983. *Working at Archaeology*. New York: Academic Press.

Bintliff, J.L. 1985. The Boeotia Survey. In *Macready, S. and Thompson, F.H. 1985*, 196-216.

Bintliff,J.L. 1991. *The "Annales" School and Archaeology*. Leicester: Leicester University Press.

Bintliff,J.L. 1992. Erosion in the Mediterranean Lands: a reconsideration of pattern, process and methodology. In *Bell, M. and Boardman, J. 1992*, 125-131.

Bintliff, J.L. 1997. Regional Survey, Demography, and the Rise of Complex Societies in the Ancient Aegean: Core-Periphery, Neo-Malthusian, and Other Interpretative Models. Boston: *Journal of Field Archaeology* 24.1, 1-38.

Bintliff, J.L. (ed). 1997a. Recent Developments in the History and Archaeology of Central Greece. Proceedings of the 6th International Boeotia Conference. Oxford: BAR (IS 666).

Bintliff, J.L., Davidson, D.A. and Grant, E.G. 1988. *Conceptual Issues on Environmental Archaeology*. Edinburgh: Edinburgh University Press.

Bintliff, J.L. and Snodgrass, A.M. 1985. The Cambridge/Bradford Boeotian Expedition: the First Four Years. Boston: *Journal of Field Archaeology* 12.2, 123-161.

Black, D. 2003. *The application of Magnetic susceptibility to the South Cadbury Environs Project*. Unpublished MA dissertation. University of Bristol.

Bradley, R. 1987. Against Objectivity. In *Gaffney, C.F. and Gaffney, V.L. 1987*, 115-119.

Brereton, S. undat. *A Survey of the results of field assessments undertaken by the Oxford Archaeological Unit and Wessex Archaeology in Southern England*. Unpublished.

Brooks, M.J., Taylor, B.E. and Grant, J.A. 1996. Carolina Bay Geoarchaeology and Holocene Landscape Evolution on the Upper Coastal Plain of South Carolina. *Geoarchaeology: An International Journal* 11.6, 481-504.

Brown, A.G. and Edwards, M.R. 1987. *Lithic Analysis and Later British Prehistory*. Oxford: BAR (162).

Brown, G., D. Field & D. McOmish. 1994. East Chisenbury Midden Complex, Wiltshire. In Fitzpatrick and Morris, 46-49.

Bryant, R.G. and Davidson, D.A. 1996. The Use of Image Analysis in the Micromorphological Study of Old Cultivated Soils: an Evaluation Based on Soils from the Island of Papa Stour, Shetland. *Journal of Archaeological Science* 23, 811-822.

Carr, C. 1991. Left in the Dust: Contextual Information in Model-Focused Archaeology. In *Kroll, E.M. and Price D.T. 1991*, 221-256.

Carrete, J-M., Keay, S. and Millett, M. 1995. *A Roman Provincial Capital and its Hinterland. The survey of the territory of Tarragona, 1985-1990*. Michigan: Journal of Roman Archaeology, Supplementary series 15

Carter, J.C. 1985. Metaponto and Croton. In *Macready, S. and Thompson, F.H. 1985*, 146-157.

Carver, M. 1996. On archaeological value. *Antiquity* 70, 45-56.

Chadwick, P. 1991. *An Assessment of Assessments: A project to assess the practice and methodology of field evaluation in rural areas*. unpublished.

Champion, T.C. (ed). 1989. *Centre and Periphery: Comparative Studies in Archaeology*. London / New York: Routledge.

Champion, T.C., Shennan, S. and Cuming, P. 1995. *Planning for the Past vol 3: Decision-making and field methods in archaeological evaluation*. University of Southampton / English Heritage.

Chapman, J; Shiel, R. and Batovic, S. 1996. *The Changing Face of Dalmatia*. Society of Antiquaries

Cherry, J.F. 1983. Frogs around the pond: perspectives in current archaeological survey projects. In *Keller, D.R. and Rupp, D,W, 1983*, 375-416.

Cherry, J.F. 1994. Regional Survey in the Aegean: The "New Wave" (and After). In *Kardulias, P.N. 1994*, 91-112.

Cherry, J.F., Davis, J.L. and Montzourani, E. 1991. *Landscape Archaeology as Longterm History: Northern Keos in Cycladic Islands*. Los Angeles: Monumenta Archgaeologica 16.

Cherry, J.F., Davis, J.L. and Montzourani, E. 1996. *The Nemea Valley Archaeological Project Internet Edition*. NVAP.

Cherry, J.F., Gamble, C. and Shennan, S. (eds). 1978. *Sampling in British Archaeology*. Oxford: BAR (50).

Chippendale, C. 1997. Editorial. *Antiquity* 71, 1-7.

Christie, N. (ed). 1995. *Settlement and Economy in Italy: 1500 BC to AD 1500*. Oxford: Oxbow (41).

Clark, R.H. and Schofield, A.J. 1991. By Experiment and Calibration: an Integrated Approach to the Archaeology of the Ploughsoil. In *Schofield, A.J. 1991*, 93-105.

Clarke, A. 1990. *Seeing Beneath the Soil*. London: Batsford.

Clarke, D.L. 1972. A Provisional Model of an Iron Age Society and its Settlement System. In *Clarke, D.L. 1972*, 801-870.

Clarke, D.L. (ed). 1972a. *Models in Archaeology*. Methuen.

Clarke, D.L. 1978. *Analytical Archaeology* (Second Edition). London: Methuen.

Coles, B. and Coles, J. 1986. *Sweet Track to Glastonbury: The Somerset Levels in Prehistory*. London: Thames & Hudson.

Coles, J. and Hall, D. 1997. The Fenland Project: from survey to management and beyond. *Antiquity* 71, 831-844.

Crawford, O.G.S. and Keiller, A. 1928. *Wessex from the Air*. Oxford.

Cunliffe, B. 1971 Hillfort's notional catchment. In *Jesson, M. and Hill, D. 1971*.

Cunliffe, B. 1994. The Danebury Environs Project, Hampshire. In *Fitzpatrick, A.P. and Morris, E.L.1994*, 38-42.

Cunliffe, B. 1995. *Danebury: An Iron Age Hillfort in Hampshire: Vol 6. A hillfort community in perspective*. York: Council for British Archaeology (102).

Cunliffe, B. 2000. *The Danebury Environs Programme: The Prehistory of a Wessex Landscape. Vol. 1: Introduction*. English Heritage and Oxford University Committee for Archaeology, Monograph 48.

Cunliffe, B. and Poole, C. 2000a. *The Danebury Environs Programme: The Prehistory of a Wessex Landscape. Vol. 2.1: Woolbury and Stockbridge Down, Stockbridge, Hants, 1989*. English Heritage and Oxford University Committee for Archaeology, Monograph 49.

Cunliffe, B. and Poole, C. 2000b. *The Danebury Environs Programme: The Prehistory of a Wessex Landscape. Vol. 2.2: Bury Hill, Upper Clatford, Hants, 1990*. English Heritage and Oxford University Committee for Archaeology, Monograph 49.

Cunliffe, B. and Poole, C. 2000c. *The Danebury Environs Programme: The Prehistory of a Wessex Landscape. Vol. 2.3: Suddern Farm, Middle Wallop, Hants, 1991 and 1996*. English Heritage and Oxford University Committee for Archaeology, Monograph 49.

Cunliffe, B. and Poole, C. 2000d. *The Danebury Environs Programme: The Prehistory of a Wessex Landscape. Vol. 2.4: New Buildings, Longstock, Hants, 1992 and Fiveways, Longstock, Hants, 1996*. English Heritage and Oxford University Committee for Archaeology, Monograph 49.

Cunliffe, B. and Poole, C. 2000e. *The Danebury Environs Programme: The Prehistory of a Wessex Landscape. Vol. 2.5: Nettlebank Copse, Wherwell, Hants, 1993*. English Heritage and Oxford University Committee for Archaeology, Monograph 49.

Cunliffe, B. and Poole, C. 2000f. *The Danebury Environs Programme: The Prehistory of a Wessex Landscape. Vol. 2.6: Houghton Down, Stockbridge, Hants, 1994*. English Heritage and Oxford University Committee for Archaeology, Monograph 49.

Cunliffe, B. and Poole, C. 2000g. *The Danebury Environs Programme: The Prehistory of a Wessex Landscape. Vol. 2.7: Windy Dido, Cholderton, Hants, 1995*. English Heritage and Oxford University Committee for Archaeology, Monograph 49.

Dall'Aglio, P.L. and Marchetti, G. 1991. Settlement Patterns and Agrarian Structures of the Roman Period in the Territory of Piacenza. In *Barker, G. and Lloyd, J. 1991*, 160-168.

D'Aubigney, A. (ed). 1990. *Opérateurs et Hypothéses pour la France*. Actes de la Table Ronde de Lons-le-Saunier.

DeGuio, A. 1995. Alto-Medio Polesine - Basso Veronese Project: From a 'Landscape Archaeology' to an 'Archaeology of the Mind'. In *Christie, N. 1995*, 13-24.

de Saussure, F. 1974. *Course in General Linguistics*. London: Fontana.

Deunert, B. 1996. *Modern Archaeology and its Reflection in the Value System of Contemporary Culture*. Oxford: BAR (IS 648)

Dietz, S; Sebaï, L.L. and Ben Hassen, H. 1995. *Africa Proconcularis: Regional Studies in the Sergemes Valley of Northern Tunisia I*. Copenhagen: National Museum of Denmark.

Dietz, S; Sebaï, L.L. and Ben Hassen, H. 1995. *Africa Proconcularis: Regional Studies in the Sergemes Valley of Northern Tunisia II*. Copenhagen: National Museum of Denmark.

Dunnell, R.C. and Dancey, W.S. 1983. The siteless survey: a regional scale data collection strategy. *Advances in Archaeological Method and Theory* (6), 267-287.

Evans, R. 1992. Erosion in England and Wales - The Present the Key to the Past. In *Bell, M. and Boardman, J. 1992*, 53-66.

Evershed, R.P., Bethell, P.H., Reynolds, P.J. and Walsh, N.J. 1997. 5☐ß-Stigmastanol and Related 5☐ß-Stanols as Biomarkers of Manuring: analysis of Modern Experimental Material and Assessment of the Archaeological Potential. *Journal of Archaeological Science* 24, 485-495.

Fellner, R.O. 1995. *Cultural Change and the Epipalaeolithic of Palestine*. Oxford: BAR (IS 599).

Fisher, P; Farrelly, C., Maddocks, A. and Ruggles, C. 1997. Spatial Analysis of Visible Areas from the Bronze Age Cairns of Mull. *Journal of Archaeological Science* 24, 581-582.

Fitzpatrick, A.P. and Morris, E.L. 1994. *The Iron Age in Wessex: Recent Work*. Salisbury: Association Française D'Etude de L'Age du Fer.

Flannery, K. (ed). 1976. *The Early Mesoamerican Village*. New York: Academic Press.

Foley, R. 1981. *Off-site archaeology and human adaptation in Eastern Africa*. Oxford: BAR (IS 97).

Foley, R. 1981a. A Model of Regional Archaeological Structure. 47, 1-17.

Forde-Johnston, J. 1976. *Hillforts of the Iron Age in England and Wales*. Liverpool.

Francovich, R. and Patterson, H. (eds). 2000. *Extracting meaning from ploughsoil assemblages*. Oxford: Oxbow Books, Archaeology of Mediterranean Landscape 5.

Gaffney, C.F. and Gaffney, V.L. 1987. *Pragmatic Archaeology: Theory in Crisis*. Oxford: BAR (167).

Gaffney, V.L. and Tingle, M. 1989. *The Maddle Farm Project: An integrated survey of Prehistoric and Roman landscapes on the Berkshire Downs*. Oxford: BAR (200).

Gater, J., Leech, R.H. and Riley, H. 1994. Later Prehistoric and Romano-British Settlement Sites in South Somerset: Some recent work. *Proceedings of the Somerset Archaeological and Natural History Society* 137, 41-58.

Gillings, M., Mattingly, D. and van Dalen J. 1999. *Geographical Information Systems and Landscape Archaeology*. Oxford: Oxbow Books, Archaeology of Mediterranean Landscapes 3.

Gregory, D. 1994. *Geographical Imaginations*. Oxford: Blackwell.

Grinsell, L.V. 1969. Somerset Barrows, Part I: West and South. *Proceedings of the Somerset Archaeological and Natural History Society* 113. (Special Supplement).

Grinsell, L.V. 1972. Somerset Barrows, Part 2: North and East. *Proceedings of the Somerset Archaeological and Natural History Society* 115, (Special Supplement).

Grinsell, L.V. 1982. *Dorset Barrows Supplement*. Dorchester: Dorset Natural History and Archaeological Society.

Haggis, D.C. 1996. Archaeological Survey at Kavousi, East Crete: Preliminary Report. *Hesperia* 65.4, 373-431.

Hall, D. 1996. *The Fenlands Project Number 10: Cambridgeshire survey. The Isle of Ely and Wisbech*. Cambridge: EastAnglian Archaeological Reports 79.

Hayes, P.P. 1985. The San Vincenzo Survey, Molise. In *Macready, S. and Thompson, F.H. 1985*, 129-135.

Hayes, P.P. and Lane, T.W. 1992. *The Fenland Project, Number 5: Lincolnshire Survey, The South-West Fens*. EastAnglian Archgaeological Reports 55.

Hazelgrove, C., Millett, M. and Smith, J. 1985. *Archaeology from the Ploughsoil*. Sheffield: University of Sheffield.

Heath, S. 1996. *The Pylos Regional Archaeological Project: Internet Edition*. http://classics.lsa.umich.edu/PRAP.html

Heffeman, K. 1996. *Liroatambo: Archaeology, history and the regional societies of Inca Cusco*. Oxford: BAR (IS 644).

Hegel, G.W.F. 1977. *The Phenomenology of Spirit*. Oxford: University Oxford Press.

Heron, C.P. and Gaffney, C.F. 1987. Archaeogeophysics and the Site: Ohm Sweet Ohm? In *Gaffney, C.F.and Gaffney, V.L. 1987*, 71-81.

Hodder, I. 1982. *The Present Past*. London: Batsford.

Hodder, I. and Malone, C. 1984. Intensive Survey of prehistoric sites in the Stilo region, Calabria. *Proceedings of the Prehistoric Society* 50, 121-150.

Hope Simpson, R. 1983. The limitations of surface survey. In *Keller, D.R. and Rupp, D.W. 1983*, 45-47.

Huggett, J. and Ryan, N. 1994. *Computer Applications and Quantitative Methods in Archaeology*. Oxford: BAR (IS 600).

Jameson, M.H. Runnels, C; van Andel, T.H. 1994. *A Greek Countryside: The Southern Argolid from Prehistory to the Present Day*. Stanford: Stanford University Press.

Jesson, M. and Hill, D. (eds). 1971. *The Iron Age and its Hillforts*. University of Southampton

Johnson, G.A. 1978. Information Sources and the development of decision-making organsiations. In *Redman, C.L. et al. 1978*, 87-112.

Kardulias, P.N. 1994. *Beyond the Site: Regional Studies in the Aegean Area*. Lanham: University Press of America.

Kardulias, P.N. 1994a. Paradigms of the Past in Greek Archaeology. In *Kardulias, P.N. 1994*, 1-23.

Keay, S.J. 1991. The Ager Tarraconensis in the Late Empire: a Model for the Economic Relationship of Town and Country in Eastern Spain? In *Barker, G. and Lloyd, J. 1991*, 79-87.

Keay, S.J. and Millett, M. 1991. Surface Survey and Site Recognition in Spain: the Ager Tarraconensis Survey and its Background. In *Schofield, A.J. 1991*, 129-139.

Keller, D.R. and Rupp, D.W. 1983. *Archaeological Survey in the Mediterranean*. Oxford: BAR (IS 155).

Kelley, K.B. 1986. *Navajo Landuse: An Ethnoarchaeological Study*. New York: Academic Press.

Kroll, E.M. and Price, D.T. (eds). 1991. The Interpretation of Archaeological Patterning. London: Plenum.

Kuhn, T.S. 1970. *The Structure of Scientific Revolutions*. Chicago: University of Chicago.

Kuna, M. 1990. Social System of the Iron Age as Reflected on the Microregional Level. In *Daubigney, A. 1990*, 227-230.

Lane, T.W. 1992. *Lincolnshire Survey: The Northern Fen Edge*. Lincoln: EastAnglian Archaeological Reports 66.

Larson, D.O., Neff, H., Graybill, D.A., Michaelsen, J. and Ambos, E. 1996. Risk, Climatic Variability, and the Study of South Western Prehistory: an evolutionary perspective. *American Antiquity* 61.2, 217-241.

Lawson, A.J. 1994. Potterne, Wiltshire. In *Fitzpatrick, A.P. and Morris, E.L. 1994*, 42-46.

Lawson, A. J. 2000. *Potterne 1982-5: Animal Husbandry in Later Prehistoric Wiltshire*. Wessex Archaeology report 17.

Leech, R. 1977. *The Upper Thames Valley in Gloucester and Wiltshire: An Archaeological Survey of the River Gravels*. Bristol: Committee for Rescue archaeology in Avon, Gloucestershire & Somerset.

Leveau, P., Trément, F., Walsh, K. and Barker, G. 1999. *Environmental reconstruction in Mediterranean archaeology*. Oxford: Oxbiow Books, Archaeology of Mediterranean Landscapes 2

Levi-Strauss, C. 1963. *Structural Anthropology*. Basic Books.

Lightfoot, K.G. 1986. Regional surveys in the Eastern United States: The strengths and weaknesses of implementing subsurface testing programmes. *American Antiquity* 51.3, 484-504.

Lightfoot, K.G. 1989. A defense of shovel-test sampling: a reply to Shott. *American Antiquity* 54.2, 413-416.

Llobera, M. 1996. Exploring the topography of mind: GIS, social space and archaeology. *Antiquity* 70, 612-622.

LLoyd, J. and Barker, G. 1981. Rural settlement in Roman Molise: problems of archaeological survey. In *Barker, G. and Hodges, R. 1981*.

Lobb, S.J. 1988. The Kennet Valley survey 1982-87: *A review of evaluation and fieldwalking methodology*. Unpublished

Lobb, S.J. and Rose, P.G. 1996. *Archaeological Survey of the Lower Kennet Valley, Berkshire*. Salisbury: Wessex Archaeology (9).

Lock, G. 2003. *Using Computers in Archaeology: towards virtual pasts*. Routledge.

Lupton, A. 1996. *Stability and Change: Socio-political development in North Mesopotamia and South-East Anatolia, 4000 - 2700 B.C.* Oxford: BAR (IS 627).

Lyall, J. and Powlesland, D. 1996. High Resolution Fluxgate Gradiometry (West Heslerton). *Internet Archaeology* 1.

MacDonald, A. 1995 All or Nothing at all? Criteria for the Analysis of Pottery from Surface Survey. In *Christie, N. 1995*. 25-29.

Macready, S. and Thompson, F.H. 1985. Archaeological Field Survey in Britain and Abroad. London: Society of Antiquaries, series VI.

Maitland Bradfield, 1973 A Natural History of Associations. London: Duckworth.

Maltby, M. 1994. Animal Exploitation in Iron Age Wessex. In *Fitzpatrick, A.P. and Morris, E.L. 1994*, 9-10.

Mattingly, D.J. 1989. Field Survey in the Libyan Valleys. *Journal of Roman Archaeology* 2, 275-280.

Mattingly, D.J. 1992. "The field survey: strategy, methodology and preliminary results". In *Ben Lazreg, N. and Mattingly, D.J. 1992*.

Mattingly, D.J. and Coccia, S. 1995. *Survey Methodology and the Site: A Roman Villa from the Rieti Survey*. In Christie, N. 1995, 31-43.

McAndrews, T.L., Albarracin-Jordan, J. and Bermann, M. 1997. Regional Settlement Patterns in the Tiwanaku Valley of Bolivia. *Journal of Field Archaeology* 24.1, 67-83.

McOmish, D. 1996. East Chisenbury: ritual and rubbish at the British Bronze Age-Iron Age transition. *Antiquity* 70, 68-76.

Millett, M. 1991. Pottery: Population or Supply Patterns? The Ager Tarraconensis Approach. In *Barker, G. and Lloyd. J. 1991*, 18-26.

Mills, N. 1981. Luni: Settlement and Landscape in the Ager Luniensis. In *Barker, G. and Hodges. R 1981*, 261-26.

Morris, E.L. 1994. The Organisation of Pottery Production in Iron Age Wessex. In *Fitzpatrick, A.P. and Morris, E.L. 1994*, 26-29.

Mueller, J.W. 1975. *Sampling in Archaeology*. Arizona: Arizona University Press

Mytum, H.C. 1988. On-site and Off-site Evidence for Changes in Subsistence Economy: Iron Age and Romano-British West Wales. In *Bintliff, J.L. et al. 1988*, 72-81.

Nance, J.D. and Ball, B.F. 1986. No surprises? The reliability and validity of Test Pit sampling. *American Antiquity* 51.3, 457-483.

Nance, J.D. and Ball, B.F. 1989. A shot in the dark: Shott's comments on Nance and Ball. *American Antiquity* 54.2, 405-412.

Nelson, B.A. 1997. Chronology and Stratigraphy at La Quemada, Zacatecas, Mexico. *Journal of Field Archaeology* 24,1, 85-109.

Newell, R.R. and Constandse-Westermann, T.S. 1996. "The use of ethnographic analyses for researching Late Palaeolithic settlement systems, settlement patterns and land use in the Northwest European Plain". In *Rowley-Conwy, P. 1996*, 372-388.

Newman, C. 1997. *Tara: An archaeological survey*. Dublin: Royal Irish Academy.

O'Connell, J.F., Hawkes, K. and Blurton-Jones, N. 1991. Distribution of Refuse-Producing Activities at Hazda Residential Base Camps. In *Kroll, E.M. and Price D.T. 1991*, 61-76.

Ørsted, P. and Sebaï, L.L. 1992. Town and Countryside in Roman Tunisia: a preliminary report on the Tuniso-Danish survey project in the Oued R'mel basin in and around ancient Sergemes. *Journal of Roman Archaeology* 5, 69-96.

Palmer, R. 1984. Danebury: An Iron Age Hillfort in Hampshire. An aerial photographic interpretation of its environs. London: RCHM(E)

Pearsall, D.M. 1996. Reconstructing Subsistence in the Lowland Tropics: A Case Study from the Jama River Valley, Manabí, Ecuador. In *Reitz et al. 1996*, 233-254.

Petropoulos, M. and Rizakis, A.D. 1994. Settlement patterns and landscape in the coastal area of Patras. Preliminary report. *Journal of Roman Archaeology* 7, 183-207.

Popper, K. 1980. The Logic of Scientific Discovery. London: Hutchinson.

Rapp, G. and Kraft, J.C. 1994. Holocene Coastal Change in Greece and Aegean Turkey. In *Kardulias, P.N. 1994*, 69-90.

Rasmussen, T. 1991. Tuscania and its Territory. In *Barker, G. and Lloyd. J. 1991*, 106-114.

Rautman, A.E. 1993. Resource variability, risk, and the structure of social networks: an example from the prehistoric southwest. *American Antiquity* 58.3, 403-424.

Rawlings, M. 1993. Romano-British Sites Observed along the Codford-Ilchester Water Pipeline. *Proceedings of the Somerset Archaeological and Natural History Society* 136, 29-60.

Redman, C.L., Berman, M.J., Curtin, E.V., Langhorne W.T., Versaggi N.M., and Wanser J.C. 1978. *Social Archaeology: Beyond Subsistence and Dating*. New York: Academic Press.

Reese, D.S. 1994. Recent work in greek Zooarchaeology. In Kardulias.

Reitz, E.J., Newsom, L.A. and Scudder, S.J. (eds). 1996. *Case Studies in Environmental Archaeology*. London: Plenum.

Rosch Heider, E. 1972. Universals in Colour Naming and Memory. *Journal of Experimental Psychology* 93.1, 10-20.

Rosch, E. 1974. Linguistic Relativity. In *Silverstein, A. 1974*.

Rosch, E., Mervis, C.B., Gray, W.D., Johnson, D.M. and Boyes-Braem, P. 1976. Basic Objects in Natural Categories. *Cognitive Psychology* 8, 382-439.

Runnels, C.N. 1994. On Lithic Studies in Greece. In *Kardulias, P.N. 1994*, 161-172.

Rowley-Conwy, P. 1996. *Hunter-Gatherer Land Use*. World Archaeology 27.3.

Russo, M. and Quitmyer, I.R. 1996. Sedentism in Coastal Populations of South Florida. In *Reitz. et al. 1996*, 215-231.

Sanchez, A., Canabate, M.L. and Lizcano, R. (eds). 1996. Phospherous Analysis at Archaeological Sites: an Optimization of the Method and Interpretation of the Results. *Archaeometry* 38.1, 151-164.

Schiffer, M.B. 1978. Methodolgical Issues in Ethnoarchaeology. In *Schiffer, M.B. 1995*, 95-106.

Schiffer, M.B. 1988. The structure of Archaeological theory. *American Antiquity* 53.3, 461-485.

Schiffer, M.B. 1995. *Behavioural Archaeology: First Principles*. Utah: University of Utah Press

Schiffer, M.B., Downing, T.E. and McCarthy, M. 1981 Waste Not, Want Not: An Ethnoarchaeological Study of Reuse in Tucson, Arizona. In *Schiffer, M.B. 1995*, 107-120.

Schiffer, M.B. and Gummerman, G.J. 1977. *Conservation Archaeology*. London: Academic Press.

Schofield, A.J. (ed). 1991. *Interpreting Artefact Scatters: contributions to ploughzone archaeology*. Oxford: Oxbow 4.

Schofield, A.J. 1991a. Artefact Distributions as Activity Areas: Examples from south-East Hampshire. In *Schofield, A.J. 1991*, 117-128.

Schofield, A.J. 1991b Interpreting artefact Scatters: an Introduction. In *Schofield, A.J. 1991*, 3-8.

Sharples, N.M. 1991. *Maiden Castle: Excavation and field survey, 1985-86*. London: English Heritage.

Shelach, G. 1998 A settlement pattern study in northeast China: results and potential contributions of western theory and methods to Chinese archaeology. *Antiquity* 72, 114-127.

Shennan, S. 1985. *Experiments in the Collection and Analysis of Archaeological Survey Data: The East Hampshire Survey*. Sheffield: John R.Collis.

Shennan, S. 1988. Quantifying Archaeology. Edinburgh: Edinburgh University Press.

Sherratt, A. 1995. *Reviving the Grand Narrative: Archaeology Reprint and Long-Term Change (Second David L. Clarke Memorial Lecture)*. Journal of European Archaeology 3 (1)

Shott, M.J. 1987. Feature Discovery and Sampling Requirements of Archaeological Evaluations. *Journal of Field Archaeology* 14, 359-371.

Shott, M.J. 1989. Shovel-testing in archaeological survey: comments on Nance and Ball, and Lightfoot. *American Antiquity* 54, 397-404.

Silverstein, A. (ed). 1974. *Human Communication: Theoretical Explorations*. Hillsdale

Simpson, I.A. 1997. Relict Properties of Anthropogenic Deep Top Soils as Indicators of Infield Management in Marwick, West Mainland, Orkney. *Journal of Archaeological Science* 24, 365-380.

Slapsak, B. 1988. Defining the Economic Space of a Typical Iron Age Hillfort: Rodik (Yugoslavia), A Case study. In *Bintliff et al. 1988*, 95-107.

Soja, E.W. 1996 *Thirdspace: Journeys to Los Angeles and other Real-and-Imagined Places*. Oxford: Blackwell.

Stein, J.K. 1986. Coring archaeological sites. *American Antiquity* 51.3, 505-527.

Steinberg, J.M. 1996. Ploughzone Sampling in Denmark: isolating and interpreting site signatures from disturbed contexts. *Antiquity* 70, 368-92.

Sutton, S.B. 1994. Settlement Patterns, Settlement Perceptions: Rethinking the Greek Village. In *Kardulias, P.N. 1994*, 313-335.

Tabor, R.J. 2002. *South Cadbury Environs Project: Interim fieldwork report, 1998-2001*. Bristol: University of Bristol.

Tabor, R.J. 2005 (forthcoming). *The landscapes, lives and deaths of Cadbury Castle: a hillfort in context*. Tempus.

Tabor, R.J. and Johnson, P.G. 2000. Sigwells, Somerset, England: regional application and interpretation of geophysical survey. *Antiquity* 74, 319-25.

Tilley, C. 1994. *A Phenomenology of Landscape: Places, Paths and Monuments*. Oxford / Providence: Berg.

Tingle, M. 1991. The Vale of the White Horse Survey. Oxford: BAR (218).

Tratman, E.K. 1970. The Glastonbury Lake Village: A Reconsideration. Proceedings of the University of Bristol Spelaeological Society 12.2

Trigger, B. 1989. A History of Archaeological Thought. Cambridge: Cambridge University Press.

van Andel, T.H. 1994. Geo-archaeology and Archaeological Science - A Personal View. In *Kardulias, P.N. 1994*, 25-44.

van der Veen, M. 1992. *Crop Husbandry Regimes: An Archaeobotanical Study of Farming in northern England*. Sheffield: J.R. Collis

van Leusen, P.M. Forthcoming. *Unbiasing the Archaeological Record*.

Voloshinov, V.N. 1973. *Marxism and the Philosophy of Language*. Seminar.

Wainwright, J. 1992. Assessing the Impact of Erosion on Semi-Arid Archaeological Sites. In *Bell, M. and Boardman, J. 1992*, 227-241.

Walters, M.R. and Kheun, D.D. 1996. The Geoarchaeology of Place: the effect of geological processes on the preservation and interpretation of the archaeological record. *American Antiquity* 61.3, 483-497.

Whimster, R. 1989. *The Emerging Past: Air Photography and the Buried Landscape*. RCHM(E).

White, R. Forthcoming. *Building an Urban Image*.

White, R. and van Leusen, P.M. Forthcoming. *Aspects of Romanization in the Wroxeter Hinterland*. TRAC'96.

Wilkinson, J. 1982. The definition of ancient manured zones by means of extensive sherd sampling techniques. *Journal of Field Archaeology* 9, 323-333.

Wittgenstein, L. 1958. *Philosophical Investigations* (third edition. Trans. Anscombe G.E.M.). Macmillan

Woodward, P.J. 1978. Flint Distribution, Ring Ditches and Bronze Age Settlement Patterns in the Great Ouse Valley: The problem, a field survey technique and some preliminary results. *Archaeological Journal* 135.

Woodward, P.J. 1991. *The South Dorset Ridgeway: Survey and excavations 1977-84*. Dorchester: DNHAS (8).

Wright, J.L., Cherry, JF; Montzourani, E. 1990. The Nemea Valley Project: A Preliminary Report. *Hesperia* 59.4, 579-659.

Yorston, R.M., Gaffney, V.L. and Reynolds P.J. 1990. Simulation of artefact movement due to cultivation. *Journal of Archaeological Science* 17.

Zeidler, JA; Pearsall, D.M. 1994. *Regional Archaeology in Northern Manabi, Ecuador. Vol 1: Environment, Cultural Chronology and Prehistoric Subsistence in the Jama Valley*. University of Pittsburgh. Memoirs in Latin American Archaeology (8).

Index

Africa Proconsularis, Tunisia 38
   environmental survey 41
   excavation in survey 59
   fieldwalking 50
   Sergemes 59
agriculture 21, 22, 24
   cultivation 22
   mixed 24, 27
   productivity 26
      ritual 26
air photography
   fieldwork based on 38
   geomorphology using 38, 39
   morphology 38
   narrative derived from 37
   phasing landscapes 38
Alcock, S 19
Alfred's Castle, Oxfordshire 38
Amboseli Survey, Kenya 48
Annales School 1, 2
   longue durée 2
Area Ratio
   definition 31
areal component matrices 32
artefact
   ceramic
      ploughzone survival 65
   comparative survival 64
   contextualisation of 30
   count 50
   diagnostic collection 50, 52
   discriminatory 64
   ecofact as 30
   ethics of collection 64
   first use *88*
   lithic dating 65
   metaphorical associations 62
   percentile analysis *79*
   sample size 49
   survival 50
   taphonomy 48
   target population 48
   taxonomy 63, *75*
   total collection
      problem of 64
artefact interpretation
   perceptual limits 61, 62
      cue validity 63
      environmental constraints 63
Aston, M 18
augering 32
Ball, B 54
Barker, G 1
Barrett, J 1, 2, 6, 28
behaviourism 1
Bell, Martin 40
Bettinger, R 2, 49
Biferno Valley, Molise 6, 23

augering 56
   data presentation 61
   excavation in survey
      environmental data 59
   exposure recording 40
   geophysical prospection 42, 46
   proton magnetometer 43
   site sampling 52
Binford, L 1, 2, 30, 61
Bintliff, J 1, 2, 6
Black, Duncan 42
Boeotia Survey, Greece 50
   site sampling 52
Bourdieu, P 2
Braudel, F 1, 2
Brean Down, Somerset
   augering 39, 56
   magnetic susceptibility 43
   test pits 40
Cadbury Castle, Somerset
   geophysical prospection 41
Carlston Annis, Kentucky
   augering 56
Carreté, J-M 19
Carver, M 2
Chapman, J 19
Cherry, J 6
   excavation in survey 59
chronology
   absolute dating *68*
   ceramic 26
   presentation of *68*
Cioffi, F *101*
Clarke, DL 1, 2, 30
   Black Box 61
Collection mosaic
   definition 31
Collection unit 33
   definition 31
   pattern 49
consciousness
   false 6
Cranborne Chase, Dorset 16
Cunliffe, B 42
   cognitive environment 25
   Danebury 29
Danebury Environs Programme, Hampshire 37
   air photography as survey 25
   environment 25
   excavation as survey 25
      environmental data 59
   hillfort territory 26
   Nettlebank Copse
      geophysics 42
   Windy Dido
      geophysics 42
Dani of New Guinea, the 62

data
  presentation
    iconic 61
    symbolic 61
data recording
  perception 61
de Guio, A *73*
demography 19
  population density 13, 21, 22, 24
  settlement density 19, 26
  settlement distribution 13, 19, 20, 21, 22
Derrida, J 1
dialectic 2
Dorset Cursus, The 16
  Cranbome Chase
    environmental data 59
Duchamp, M
  toilet 4
Dunnell, R 1
East Hampshire, Britain 50
economy 21, 22
  agriculture 17
  division of labour 23, 24
  exchange 25
  industry 13, 22, 24
  specialisation 13, 22, 27
  transition to farming 23
empiricism 1
environment
  reconstruction 10, 12, 15, 16, 17, 21, 32, 49
environmental data
  excavation in survey 59, *92*
  erosion 16, 18, 40
  prehistoric agriculture 40
ethnography 2
evidence
  negative
    problems of 57, *96*
evolutionary 1
experimental archaeology
  Butser Ancient Farm *66*
  ethnoarchaeology *86*
Fenland Project, The
  exposure recording 40
field systems
  implementation of 25
fieldwalking
  performance variation *71*
  rate of progress 49, 50
  resolution *70*
  site sampling
    encounter 52
  technique 49, 50
    clicker 50
Foley, R 2, 6, 18, 48, 49
  model-based narrative 27
Foucault, M 1
Gadamer, H 7
Gaffney, V 13, 16
Geographical Information Systems (GIS) 31, 37, 38
  Wroxeter *30*

geomorphology 10, 18
  erosion 25
geophsyical prospection
  gradiometer
    Bartington 42
    depth range 46
    Geoscan 42
    sampling frequency 43
    West Heslerton, Yorkshire 46
    Wroxeter, Shropshire 46
  Wroxeter *30*
Gerrard, C 18
  ethics of collection 64
Giddens, A 1
Great Ouse Valley, Bedfordshire 19
Gregory, D 2
Grid
  definition 31
Grinsell, L 38
Ground Penetrating Radar (GPR) 42
Habermas, J 1
Hegel, G 1, 2, 28
  dialectic 2
Hodder, I 2, 20, 49
hunter-gatherer 21
ideology
  belief 26
  weltanschaung 7
interaction
  aggression 20, 22, 25
  trade 19, 20, 21, 22, 24, 27
Jama Valley, Ecuador *68*
Kavousi, Crete *72*
Kuhn, T 1
labour
  volunteer *64*
land resource
  selection 13, 20
landscape
  continuity 6
  discontinuity 17
  division of 25
  human activity in 29
  tradition 6
Levi-Strauss, C 2
Libyan Valleys Survey 39
  data presentation 61
  environmental survey 41
Lightfoot, K 54
Line
  definition 31
linguistic determinism
  categorisation 62
Lloyd, J 1
longue durée 2
Maddle Farm, Berkshire 14, 16, 33
  fieldwalking 50
  test pits
    sub surface survey 57
Malone, C 49
manuring scatters 14, 17

map
  regressive analysis of 37
  use of air photographs 37
material archive 18
  selective 18
Mesh
  definition 31
metalwork
  hoards 18
methodology
  problem-solving 28
middens 18
Millett, M 56
Miwa, Japan 30, 39
models
  agricultural production 77
  arenas of social power 13, 19, 28
  case-specific 7
  communal membership of property model 13
  Core-Periphery 7
  cyclic intensification-deintensificadon model 13, 14
  digital terrain 38
  ethnographic 7, 10, 30, 49, 61
  in practice 61
  Land Use Capability 7, 13, 14
  narrative 10
  prejudice 7
  problem-solving 13, 16, 18
  Thiessen polygon *82*
Musson, C 41
Nance, J 54
narrative
  class A 4, 6
  class B 4, 6
  class C 4, 6
  class D 4, 6, 10
  class E 4, 6, 10, 19
  class F 4, 6, 7, 13, 15, 19
  class G 4, 7, 15, 17, 19, 29
  economic 6
  feedback loops 6
  Grand 2
  political 6
  sequence 4
  social 19
  structure 6
National Monuments Record (NMR) 37, 48
Nemea Valley Project *67*
Neo-Thermal Dalmatia Project 19
  site sampling 52
North Keos, Cyclades 20
  site sampling 52
Owens Valley, USA 49
Palmer, R 25, 37, *91*, 96
paradigms
  agency 16, 28
  analytical 1, 61
  Annales-school 1
  behavioural 1
  empiricism 1
  evolutionary 1, 12, 28

  phenomenological 28
  positivism 1-3
  problem-solving 7, 15
  processual 1, 30
  regional off-site 7
  spatial 13
  sub-paradigm 7
perception
  classification *62*
  fieldwalkers *65*
Pitman, B 2
ploughzone sampling 46
  homogenous distribution 46
political
  kinship 17
  urbanisation 24
Popper, K 4
  pseudo-science 4
post-structuralism 2
processualism 1-3
proton magnetometer 41
recording forms
  field 33
recording forms, field 33
region
  concept of 30
resistivity
  seasonal variation 46
resource distribution 32
Ricoeur, P 7
ritual 17, 24
  dispoal of the dead 26
  significance 17
Rosch, E
  cue validity 63
  lingguistic relativism 63
  perception 62
Runnels, C
  lithic analysis 65
sample
  sufficient *10*
Sample point
  definition 31
sampling 33
  discovery probability 52
  evaluation 52
  frequency 49
  from an assemblage *77*
  grab 52
  judgement 49
  site
    iron cross 54
Sapir, E
  linguistic determinism 62
Schadla Hall, R 57
Schiffer, M 1, 2
Schofield, AJ 18, 19
Seamer Carr, Yorkshire
  augering 57
  test pitting 57
Segment

definition 31
settlement 26
Shanks, M 20
Shapwick Project, Somerset
   fieldwalking 50
   shovel pitting 54
Shennan, S 2
Sherratt, A 2
Shott, M 54
shovel pitting 52
   definition 48
   iron cross 54
   regular interval 54
site
   conceptual problem of 31
   extent problem 54
   location preference *81*
   signature 13
Sites and Monuments Records (SMR) 37, 48
soil conductivity meter 41
Soja, E 2
South Cadbury Environs Project 3, *73*
   Fieldwalking 50
   fieldwalking 50
   geology *96*
   gradiometer survey 43
   land use *98*
   test pits 40, 57
   topography *96*
South Dorset Ridgeway 15, 19, 49
Southern Argolid, Pelopennese 6, 21
   fieldwalking 50
   site definition 52
   site sampling 52
space
   definition of 7
statistical analysis *76*
Stilo, Calabria 49
strata 32
   biomass, use of 10
   ecozone 49
Stratum
   definition 31
structuralism 2
sub-paradigm 7
surface collection
   experiment 49
   sampling strategy 49
survey
   history of 30, 32
   strata 10, 32
Survey Area
   definition 31

taphonomy 7
   geomorphology 7
Tara, Eire 38
   geophysics 42
Tarragona, Catalunya 18
   El Vilar 19
   fieldwalking 50
   resistivity survey 46
   site 3.19 19
   stratified artefacts 59
Taylor, J
   ploughzone ceramics 65
technique selection 60
test pit (TP)
   deep soils 57
   definition 48
   evaluation using 57
   sampling frequency 57
Thorpe, N 52
Thy, Jutland 56
Tilley, C 1, 20, 28
Tingle, M 13, 16
topsoil sampling 46
Transect
   definition 31
Trigger, B 1
van Andel, T 41
van der Veen, M 41
variables 7, 36
   ground conditions 10, 36
   visibility 36
visibility 36
Vita Finzi, C 30
Volnshinov, V
   theme in meaning *101*
Wessex Hillforts Project 42
   magnetic survey 42
   magnetometer survey 42
West Heslerton, Yorkshire
   gradiometer survey 46
Whimster, R 37
Whorf, B
   linguistic determinism 62
Wittgenstein, L 61
   duck-rabbit 62
Woodward, P 18, 19
Wroxeter Hinterland Project
   ground penetrating radar 42
   proton magnetometer 43
   resistivity survey 46
Young, P 46
Zinancantecos of Mexico, The
   colour perception 62

www.ingramcontent.com/pod-product-compliance
Ingram Content Group UK Ltd.
Pitfield, Milton Keynes, MK11 3LW, UK
UKHW060200240426
12048UKWH00029B/1666